Applications of
Affine and Weyl Geometry

Synthesis Lectures on Mathematics and Statistics

Editor
Steven G. Krantz, *Washington University, St. Louis*

Applications of Affine and Weyl Geometry
Eduardo García-Río, Peter Gilkey, Stana Nikčević, and Ramón Vázquez-Lorenzo
2013

Essentials of Applied Mathematics for Engineers and Scientists, Second Edition
Robert G. Watts
2012

Chaotic Maps: Dynamics, Fractals, and Rapid Fluctuations
Goong Chen and Yu Huang
2011

Matrices in Engineering Problems
Marvin J. Tobias
2011

The Integral: A Crux for Analysis
Steven G. Krantz
2011

Statistics is Easy! Second Edition
Dennis Shasha and Manda Wilson
2010

Lectures on Financial Mathematics: Discrete Asset Pricing
Greg Anderson and Alec N. Kercheval
2010

Jordan Canonical Form: Theory and Practice
Steven H. Weintraub
2009

The Geometry of Walker Manifolds
Miguel Brozos-Vázquez, Eduardo García-Río, Peter Gilkey, Stana Nikcevic, and Ramón Vázquez-Lorenzo
2009

An Introduction to Multivariable Mathematics
Leon Simon
2008

Jordan Canonical Form: Application to Differential Equations
Steven H. Weintraub
2008

Statistics is Easy!
Dennis Shasha and Manda Wilson
2008

A Gyrovector Space Approach to Hyperbolic Geometry
Abraham Albert Ungar
2008

Applications of Affine and Weyl Geometry

Eduardo García-Río, Peter Gilkey, Stana Nikˇceviˊc, and Ramón Vázquez-Lorenzo

ISBN: 978-3-031-01277-8   paperback
ISBN: 978-3-031-02405-4   ebook

DOI 10.1007/978-3-031-02405-4

A Publication in the Springer series
SYNTHESIS LECTURES ON MATHEMATICS AND STATISTICS

Lecture #13
Series Editor: Steven G. Krantz, *Washington University, St. Louis*
Series ISSN
Synthesis Lectures on Mathematics and Statistics
Print 1938-1743   Electronic 1938-1751

Applications of
Affine and Weyl Geometry

Eduardo García-Río
University of Santiago de Compostela

Peter Gilkey
University of Oregon

Stana Nikčević
Mathematical Institute, Sanu, Serbia

Ramón Vázquez-Lorenzo
University of Santiago de Compostela

SYNTHESIS LECTURES ON MATHEMATICS AND STATISTICS #13

ABSTRACT

Pseudo-Riemannian geometry is, to a large extent, the study of the Levi-Civita connection, which is the unique torsion-free connection compatible with the metric structure. There are, however, other affine connections which arise in different contexts, such as conformal geometry, contact structures, Weyl structures, and almost Hermitian geometry. In this book, we reverse this point of view and instead associate an auxiliary pseudo-Riemannian structure of neutral signature to certain affine connections and use this correspondence to study both geometries. We examine Walker structures, Riemannian extensions, and Kähler–Weyl geometry from this viewpoint. This book is intended to be accessible to mathematicians who are not expert in the subject and to students with a basic grounding in differential geometry. Consequently, the first chapter contains a comprehensive introduction to the basic results and definitions we shall need—proofs are included of many of these results to make it as self-contained as possible. Para-complex geometry plays an important role throughout the book and consequently is treated carefully in various chapters, as is the representation theory underlying various results. It is a feature of this book that, rather than as regarding para-complex geometry as an adjunct to complex geometry, instead, we shall often introduce the para-complex concepts first and only later pass to the complex setting.

The second and third chapters are devoted to the study of various kinds of Riemannian extensions that associate to an affine structure on a manifold a corresponding metric of neutral signature on its cotangent bundle. These play a role in various questions involving the spectral geometry of the curvature operator and homogeneous connections on surfaces. The fourth chapter deals with Kähler–Weyl geometry, which lies, in a certain sense, midway between affine geometry and Kähler geometry. Another feature of the book is that we have tried wherever possible to find the original references in the subject for possible historical interest. Thus, we have cited the seminal papers of Levi-Civita, Ricci, Schouten, and Weyl, to name but a few exemplars. We have also given different proofs of various results than those that are given in the literature, to take advantage of the unified treatment of the area given herein.

KEYWORDS

curvature decomposition, deformed Riemannian extension, Kähler–Weyl geometry, modified Riemannian extension, Riemannian extension, spectral geometry of the curvature operator

This book is dedicated to
Carmen, Emily, George, Hugo, Luis, Manuel, Montse, and Susana.

Contents

Preface

The fundamental theorem of pseudo-Riemannian geometry associates to each pseudo-Riemannian metric g a unique affine connection,

$$\nabla = {}^g\nabla \,,$$

called the Levi-Civita connection (we refer to Levi-Civita [151] and to Ricci and Levi-Civita [188]), and pseudo-Riemannian geometry focuses, to a large extent, on the geometry of this connection. There are, however, other natural connections that play a role when considering different geometric structures, such as almost Hermitian structures, almost contact structures, and Weyl structures. Affine connections also arise naturally in conformal geometry. In all these cases, the affine connections under consideration are adapted to the structures under investigation. One can reverse this point of view and associate an auxiliary pseudo-Riemannian structure to a given affine connection and then use this correspondence to examine the geometry of both objects. As an exemplar, the Riemannian extension is a natural, neutral signature, pseudo-Riemannian metric on the cotangent bundle of an underlying affine manifold. This construction, which goes back to the work of Patterson and Walker [177], has played an important role in many investigations.

This book examines a number of different areas of differential geometry which are related to affine differential geometry—Walker structures, Riemannian extensions, and (para)-Kähler–Weyl geometry. It is intended to be accessible to graduate students who have had a basic course in differential geometry, as well as to mathematicians who are not necessarily experts in the subject. For that reason, Chapter 1 contains a basic introduction to the matters under consideration. The Lie derivative and bracket, connections, Kähler geometry, and curvature are discussed. The notion of a curvature model is treated and the basic decomposition theorems of Singer–Thorpe [194], Higa [125, 126], and Tricerri–Vanhecke [200] are given, not only in the Riemannian, but in the pseudo-Riemannian and in the para-complex settings as well. Almost para-Hermitian and almost Hermitian structures are presented both in the Riemannian and pseudo-Riemannian categories. The Gray–Hervella [118] classification of almost Hermitian structures is extended to the almost para-Hermitian and to the almost pseudo-Hermitian contexts. The geometry of the cotangent bundle is outlined (tautological 1-form, evaluation map, complete lift) and the various natural metrics on the cotangent bundle (Riemannian extensions, deformed Riemannian extensions, modified Riemannian extensions) are defined – these will be examined in further detail in Chapter 2 and Chapter 3. A short introduction to Walker geometry and recurrent curvature is included. Self-dual Walker metrics are discussed and it is shown that any such metric is locally isometric to the metric of a Riemannian extension. The Jacobi operator,

$$\mathcal{J}(x) : y \to \mathcal{R}(y, x)x \,,$$

is introduced and classical results concerning symmetric spaces, spaces of constant sectional curvature, and constant holomorphic sectional curvature are presented using Jacobi vector fields. Chapter 1 concludes with a brief review of the spectral geometry of the curvature operator. Complete proofs of a number of results are presented to keep the treatment as self-contained as possible.

Chapter 2 examines the geometry of Riemannian extensions in more detail and extends the discussion of Chapter 1 in that regard. Riemannian extensions form a natural family of examples and provide a first link between affine and pseudo-Riemannian geometry. The relevant facts concerning the metric of the classical Riemannian extension are developed in some detail. Riemannian extensions are then used to study the spectral geometry of the curvature tensor in the affine setting; affine Osserman surfaces and affine Ivanov–Petrova surfaces are examined in some detail. Chapter 2 concludes with a fairly lengthy treatment of homogeneous affine connections. Homogeneous connections on surfaces that are not the Levi-Civita connection of a metric of constant curvature form two natural classes which are not disjoint, and the intersection of the two non-metric classes is studied using the corresponding Riemannian extensions.

Chapter 3 presents generalizations of Riemannian extensions. While Riemannian extensions are useful in constructing self-dual Ricci flat metrics, the modified Riemannian extensions are a source of non-Ricci flat Einstein metrics. Four-dimensional geometry is explored in some detail. A new approach, based on the generalized Goldberg–Sachs theorem, is used to obtain some previously known results on the classification of 4-dimensional Osserman metrics from this point of view. The usefulness of modified Riemannian extensions is made clear when discussing 4-dimensional Osserman metrics whose Jacobi operators have non-degenerate Jordan normal form. Para-Kähler manifolds of constant para-holomorphic sectional curvature are treated in this fashion, as are a variety of higher-dimensional Osserman metrics – it is shown that any such manifold is locally isometric to the modified Riemannian extension metric of a flat affine manifold. By considering non-flat affine Osserman connections, one obtains a family of *deformed* para-Kähler metrics with the same spectrum of the Jacobi operator. This provides a useful family of Osserman metrics whose Jacobi operators have non-trivial Jordan normal form. The related property of being (semi) para-complex Osserman is also considered, and it is shown that any modified Riemannian extension is a semi para-complex Osserman metric. Here, the curvature identities induced by the almost para-complex structure play an essential role; note that the para-holomorphic sectional curvature does not necessarily determine the curvature in the general setting.

Chapter 4 treats (para)-Kähler–Weyl geometry. Classical Weyl geometry is, in a certain sense, midway between affine and Riemannian geometry, and (para)-Kähler–Weyl geometry is a natural generalization of Kähler geometry. Results are presented both in the complex and para-complex settings. Only the 4-dimensional setting is of interest in this context – any (para)-Kähler–Weyl structure is trivial in dimension $m \geq 6$. By contrast, any para-Hermitian manifold or any pseudo-Hermitian manifold admits a unique (para)-Kähler–Weyl structure if m=4. The alternating Ricci tensor ρ_a carries essential information about the structure in dimension four – the (para)-Kähler–Weyl structure is trivial if and only if $\rho_a = 0$. All the possible algebraic possibilities for the values of ρ_a

can be realized by left-invariant structures on 4-dimensional Lie groups for Hermitian manifolds or para-Hermitian manifolds – the case of pseudo-Hermitian manifolds of signature (2, 2) is excluded, as quite different techniques would be needed to study that case that are tangential to the thrust of our main discussion. This result is used to examine the space of algebraic (para)-Kähler–Weyl curvature tensors as a module over the appropriate structure groups and to show that any algebraic possibility can be realized geometrically.

Eduardo García-Río, Peter Gilkey, Stana Nikčević, and Ramón Vázquez-Lorenzo
April 2013

Acknowledgments

The research of all of the authors was partially supported by Project MTM2009-07756 (Spain); the research of P. Gilkey and S. Nikčević was partially supported by Project 174012 (Srbija). The authors are very grateful to Esteban Calviño-Louzao for assistance in proofreading.

Eduardo García-Río, Peter Gilkey, Stana Nikčević, and Ramón Vázquez-Lorenzo
April 2013

CHAPTER 1

Basic Notions and Concepts

In this chapter, we establish notation and introduce the fundamental concepts that we will be dealing with. Some of the basic theory of manifolds that is needed is presented in Section 1.1. Section 1.2 introduces the notion of a connection and the associated curvature operator on a general vector bundle. Of particular interest is the case when the bundle in question is the tangent bundle. The torsion tensor $\mathcal{T}$ is introduced; a connection is called affine if $\mathcal{T}$ vanishes. The Levi-Civita connection is the unique torsion-free Riemannian connection; the notion of a Weyl structure is midway between that of an affine structure and the Levi-Civita connection. In Section 1.3, curvature models in the real setting are examined. Basic geometric realizability results are given in Theorem 1.12. In Section 1.4, we pass to the complex and to the para-complex settings. A formalism is introduced that enables us to treat both contexts in a parallel notation. Section 1.5 deals with curvature decompositions and basic representation theory. The Singer–Thorpe [194], Higa [125, 126], and Tricerri–Vanhecke [200] decompositions are given for the curvature tensor in the Riemannian, the Weyl, and the Hermitian settings, respectively. Section 1.6 introduces Walker structures and Riemannian extensions. It treats the holonomy group, parallel plane fields, and Walker coordinates. Section 1.7 deals with metrics on the cotangent bundle: Riemannian extensions, deformed Riemannian extensions, and modified Riemannian extensions. These notions will be analyzed in more detail subsequently. Section 1.8 discusses the Hodge $\star$ operator, self-duality, and self-dual Walker metrics. Section 1.9 is concerned with recurrent curvature. Section 1.10 deals with constant curvature and introduces the notions of a Jacobi vector field, of a symmetric space, of constant sectional curvature, and of constant holomorphic sectional curvature. Section 1.11 provides an introduction to the spectral geometry of the curvature. Osserman manifolds, higher order Osserman manifolds, and the skew-symmetric curvature operator are introduced.

1.1 BASIC MANIFOLD THEORY

Let M be a connected smooth manifold of dimension m—for further information see Kobayashi and Nomizu [140]. If $\vec{x} = (x^1, \ldots, x^m)$ is a system of local coordinates on M, set

$$\partial_{x_i} := \frac{\partial}{\partial x^i};$$

$\{\partial_{x_1}, \ldots, \partial_{x_m}\}$ and $\{dx^1, \ldots, dx^m\}$ provide local *coordinate frames* for the *tangent bundle* TM and for the *cotangent bundle* T^*M. We shall also, sometimes, use the notation ∂_i for ∂_{x_i} when no confusion is likely to result. We shall, for the most part, follow the convention of writing indices up that refer to *contravariant tensors* (coordinate systems, differential forms, etc.) and indices down that refer to *covariant tensors* (partial differential operators). An exception will occur to this convention presently, when we discuss the canonical coordinates on the cotangent bundle (see Equation (1.1.e)). Thus, for example, if $I = \{1 \le i_1 < \cdots < i_p \le m\}$ is a multi-index, let $dx^I := dx^{i_1} \wedge \cdots \wedge dx^{i_p}$; the $\{dx^I\}_{|I|=p}$ provide a local frame for the bundle of *p-forms* $\Lambda^p M$. Let $S^2(M)$ be the bundle of *symmetric 2-cotensors* on M. If ξ and η are 1-forms, the *symmetric tensor product* of ξ with η is defined by setting:

$$\xi \circ \eta := \tfrac{1}{2}\{\xi \otimes \eta + \eta \otimes \xi\}. \tag{1.1.a}$$

The $\{dx^i \circ dx^j\}_{1 \le i \le j \le m}$ form a local frame for $S^2(M)$. Let $d : C^\infty(\Lambda^p M) \to C^\infty(\Lambda^{p+1} M)$ be the *exterior derivative*. Here, and henceforth, we will adopt the *Einstein convention* and sum over repeated indices. Thus, for example, we may express:

$$df = \partial_{x_i} f \cdot dx^i \text{ for } f \in C^\infty(M) \text{ and } d\left(\sum_I f_I dx^I\right) = \sum_I df_I \wedge dx^I.$$

Since $d^2 = 0$, the *de Rham cohomology groups* may be defined by setting:

$$H^p_{\mathrm{DeR}}(M) := \frac{\ker\left\{d : C^\infty(\Lambda^p M) \to C^\infty(\Lambda^{p+1} M)\right\}}{\mathrm{Range}\left\{d : C^\infty(\Lambda^{p-1} M) \to C^\infty(\Lambda^p M)\right\}}.$$

These groups are isomorphic to the ordinary topological cohomology groups (see de Rham [185]), so this provides an important link between differential geometry and topology. Let

$$\delta : C^\infty(\Lambda^p M) \leftarrow C^\infty(\Lambda^{p+1} M) \tag{1.1.b}$$

be the dual map, the *coderivative*. One forms the *Laplacian*

$$\Delta := d\delta + \delta d : C^\infty(\Lambda^p M) \to C^\infty(\Lambda^p M).$$

The Hodge–de Rham theorem [128] identifies $\ker\{\Delta\}$ with $H^p_{\mathrm{DeR}}(M)$ if M is closed. There are suitable extensions of this result to the context of manifolds with boundary—see, for example, the discussion in Gilkey [95].

SYMPLECTIC AND CONTACT STRUCTURES

If $m = 2\bar{m}$ is even, then a *symplectic structure* on M is a 2-form Ω such that

$$d\Omega = 0 \text{ and } \Omega^{\bar{m}} \ne 0.$$

The pair (M, Ω) is then called a *symplectic manifold*. As we shall see presently in Equation (1.1.f), the cotangent bundle always admits a canonical symplectic structure. Also, any (para)-Kähler metric defines a canonical symplectic structure—see Equation (1.5.c).

If (M, Ω) is a symplectic manifold of dimension $m = 2\bar{m}$, then by Darboux's theorem [61] there always exist local coordinates (p^i, q^i) so that

$$\Omega = \sum_{i=1}^{\bar{m}} dp^i \wedge dq^i .$$

Thus, there are no local invariants in symplectic geometry. The structure has its origin in classical mechanics—the phase space in Hamiltonian mechanics has the structure of a symplectic manifold where the p coordinates describe the momenta and the q coordinates describe the position of a particle. For further details, we refer to de Gosson [116], to McDuff and Salamon [154], and to Weinstein [205].

If $m = 2\bar{m} + 1$ is odd, then the natural analogue of symplectic geometry is *contact geometry*. Let η be a smooth 1-form on M. We say that η is a *contact 1-form* and that the pair (M, η) is a *contact manifold* if

$$\eta \wedge (d\eta)^{\bar{m}} \neq 0 .$$

The *Reeb vector field* ξ is characterized by the property that

$$\eta(\xi) = 1 \text{ and } d\eta(\xi, X) = 0 \text{ for all } X \in C^\infty(TM) .$$

We refer to Geiges [93] for further details.

THE LIE DERIVATIVE AND THE LIE BRACKET

Let $X \in C^\infty(TM)$ be a smooth vector field on M. A *flow curve* for X starting at the point P is the solution to the equation:

$$\dot{\alpha}(t) = X(\alpha(t)) \text{ with } \alpha(0) = P .$$

The smooth *1-parameter flow* associated to X is the family of local diffeomorphisms comprising the collection of flow curves. More precisely, this is a family of local diffeomorphisms φ_t^X of M so that the curves $\alpha : t \to \varphi_t^X(P)$ are flow curves:

$$\dot{\varphi}_t^X(P) = X(\varphi_t^X(P)) \text{ with } \varphi_0^X(P) = P \text{ for all } P \in M .$$

We note that

$$\varphi_t^X \varphi_s^X = \varphi_{s+t}^X .$$

If Ψ is a *tensor field* on M (i.e., $\Psi \in \otimes^p T^*M \otimes^q TM$ for some (p, q)), then we can form the tensor $(\varphi_t^X)^*\Psi$. The *Lie derivative* is then defined by setting:

$$\mathcal{L}_X \Psi := \partial_t \left\{ (\varphi_t^X)^* \Psi \right\} \Big|_{t=0} .$$

Let P be a point of M. If $X(P) \neq 0$, we can choose a system of local coordinates $(x^1, \ldots, x^m)$ so that $X = \partial_{x_1}$ and so that the flow is given locally by

$$\varphi_t^X(x^1, \ldots, x^m) = (x^1 + t, x^2, \ldots, x^m) .$$

If we express

$$\Psi = a_{i_1,\dots,i_p}^{j_1,\dots,j_q} \cdot dx^{i_1} \otimes \cdots \otimes dx^{i_p} \otimes \partial_{x_{j_1}} \otimes \cdots \otimes \partial_{x_{j_q}} \,,$$

we then have

$$\mathcal{L}_X \Psi = \left(\partial_{x_1} a_{i_1,\dots,i_p}^{j_1,\dots,j_q} \right) \cdot dx^{i_1} \otimes \cdots \otimes dx^{i_p} \otimes \partial_{x_{j_1}} \otimes \cdots \otimes \partial_{x_{j_q}} \,.$$

Let $[X, Y]$ denote the *Lie bracket* of two vector fields $X, Y \in C^\infty(TM)$. This is characterized by the identity:

$$[X, Y]f = X(Y(f)) - Y(X(f)) \text{ for all } f \in C^\infty(M)\,.$$

We can express the Lie bracket in a system of local coordinates as follows. If $X = a^i \partial_{x_i}$ and if $Y = b^j \partial_{x_j}$ are smooth vector fields on M, then:

$$[X, Y] = \left\{ a^i \partial_{x_i}(b^j) - b^i \partial_{x_i}(a^j) \right\} \partial_{x_j}\,.$$

One has the *Jacobi identity*:

$$[[X, Y], Z] + [[Y, Z], X] + [[Z, X], Y] = 0\,.$$

Let $\ll \cdot, \cdot \gg$ denote the natural *bilinear pairing* between a vector bundle $\mathbb{V}$ and the dual bundle $\mathbb{V}^*$. It is characterized on the tangent bundle by the identity:

$$\ll \partial_{x_i}, dx^j \gg = \delta_i^j \text{ where } \delta_i^j := \left\{ \begin{array}{ll} 1 & \text{if } i = j \\ 0 & \text{if } i \neq j \end{array} \right\}$$

is the *Kronecker symbol*. The Lie bracket and the exterior derivative are related by the formula:

$$d\omega(X, Y) = X \ll Y, \omega \gg - Y \ll X, \omega \gg - \ll [X, Y], \omega \gg \,. \tag{1.1.c}$$

Similarly, the Lie bracket and the Lie derivative are related by the formula:

$$[X, Y] = \mathcal{L}_X Y\,.$$

The Lie bracket and the flows are also related (see Spivak [196, Chapter 5 vol. 1]). The commutator flow

$$c(t; P) := \varphi_{-\sqrt{t}}^Y (\varphi_{-\sqrt{t}}^X (\varphi_{\sqrt{t}}^Y (\varphi_{\sqrt{t}}^X (P)))) \text{ satisfies } \dot{c}(t; P)|_{t=0} = [X, Y](P)\,.$$

GEOMETRY OF THE COTANGENT BUNDLE

We refer to Jost [137] and to Yano and Ishihara [215] for further details concerning this material. Let $\sigma : T^*M \to M$ be the *natural projection* from the cotangent bundle to the base manifold. We may express any point $\tilde{P}$ of T^*M in the form $\tilde{P} = (P, \omega)$ where $P := \sigma(\tilde{P})$ belongs to M and where ω belongs to T_P^*M. Let $\bar{m} = \dim(M)$ and let

$$\vec{x} = (x^1, \dots, x^{\bar{m}})$$

be a system of local coordinates on an open neighborhood U of $P \in M$. Expanding

$$\omega = x_{i'} dx^i \tag{1.1.d}$$

then defines a system of local coordinates on $\tilde{U} := \sigma^{-1} U \subset T^* M$ of the form

$$(x^1, \ldots, x^{\bar{m}}; x_{1'}, \ldots, x_{\bar{m}'}). \tag{1.1.e}$$

The 1-form ω of Equation (1.1.d) is often referred to as the *tautological 1-form*, although, it is also denoted as the *Poincaré 1-form* or the *Liouville 1-form*. The dual coordinates are written with the index down, as they transform covariantly rather than contravariantly and this permits us to retain the formalism of summing over repeated indices. Thus the canonical *symplectic structure* on the cotangent bundle is given in terms of a system of local coordinates by:

$$\Omega := d\omega = dx_{i'} \wedge dx^i. \tag{1.1.f}$$

We will show presently that Ω is invariantly defined and is independent of the particular coordinate system chosen for its evaluation by giving an invariant characterization in Equation (1.1.h). Note that Ω is a closed differential form with $\Omega^{\bar{m}} \neq 0$.

Let $X \in C^\infty(TM)$ be a smooth vector field on M. The *evaluation map* ιX is a smooth function on the cotangent bundle $T^* M$ which is defined by the identity:

$$\iota X(P, \omega) = \omega(X_P).$$

We may express $X = X^i \partial_{x_i}$ to define the coefficients $X^i = \ll X, dx^i \gg$. We then have that

$$\iota X(x^i, x_{i'}) = x_{i'} X^i = x_{i'} \ll X, dx^i \gg .$$

The special significance of the evaluation map comes from the fact that vector fields on $T^* M$ are characterized by their action on the evaluation maps ιX (see Yano and Ishihara [215]). More precisely:

Lemma 1.1 *Suppose that $\tilde{Y}$ and $\tilde{Z}$ are smooth vector fields on the cotangent bundle $T^* M$. Suppose that $\tilde{Y}(\iota X) = \tilde{Z}(\iota X)$ for all smooth vector fields X on M. Then $\tilde{Y} = \tilde{Z}$.*

Proof. Let $\tilde{Y} = a^i(\vec{x}, \vec{x}') \partial_{x_i} + b_{i'}(\vec{x}, \vec{x}') \partial_{x_{i'}}$ be a smooth vector field on $T^* M$ such that $\tilde{Y}(\iota X) = 0$ for all smooth vector fields on M. We must show this implies that $\tilde{Y} = 0$. Let $X = X^j(\vec{x}) \partial_{x_j}$. Since $\iota X = x_{i'} X^i$, we have:

$$0 = \tilde{Y}(\iota X) = x_{j'} a^i(\vec{x}, \vec{x}')(\partial_{x_i} X^j)(\vec{x}) + b_{i'}(\vec{x}, \vec{x}') X^i(\vec{x}).$$

Fix j and do not sum. If we take $X = \partial_{x_j}$, then $X^i = \delta^i_j$ so $\iota X = x_{j'}$ and

$$0 = \tilde{Y}(\iota X) = b_{j'}(\vec{x}, \vec{x}') \text{ so } b_{j'} = 0 \text{ and } \tilde{Y} = a^i(\vec{x}, \vec{x}') \partial_{x_i}.$$

If we take $X = x^j \partial_{x_j}$, then we have similarly

$$0 = \tilde{Y}(\iota X) = x_{j'} a^j(\vec{x}, \vec{x}') .$$

Thus $a^j(\vec{x}, \vec{x}') = 0$ when $x_{j'} \neq 0$. Since the functions $a^j(\vec{x}, \vec{x}')$ are smooth, this implies a^j vanishes identically and hence $\tilde{Y} = 0$ as desired. $\qquad\square$

Let X be a smooth vector field on M. The *complete lift* X^C is the vector field on the cotangent bundle T^*M characterized via Lemma 1.1 by the identity:

$$X^C(\iota Z) = \iota[X, Z] \text{ for all smooth vector fields } Z \text{ on } M.$$

Lemma 1.2   *Let (P, ω) belong to $T^*M - \{0\}$, i.e., $\omega \neq 0$. Then the tangent space $T_{(P,\omega)}T^*M$ is spanned by the complete lifts of all the smooth vector fields on M.*

Proof. We first compute the complete lift in a system of local coordinates. Let

$$X = X^j(\vec{x})\partial_{x_j} \text{ and } X^C = a^j(\vec{x}, \vec{x}')\partial_{x_j} + b_{j'}(\vec{x}, \vec{x}')\partial_{x_{j'}} .$$

Let $Z = Z^i(\vec{x})\partial_{x_i}$. Then

$$
\begin{aligned}
X^C(\iota Z) &= X^C(x_{i'}Z^i) = x_{i'}a^j(\vec{x}, \vec{x}')(\partial_{x_j}Z^i)(\vec{x}) + b_{j'}(\vec{x}, \vec{x}')Z^j(\vec{x}) \\
&= \iota\{(X^j\partial_{x_j}Z^i - Z^j\partial_{x_j}X^i)\partial_{x_i}\} \\
&= x_{i'}(X^j(\vec{x})(\partial_{x_j}Z^i)(\vec{x}) - Z^j(\vec{x})(\partial_{x_j}X^i)(\vec{x})) .
\end{aligned}
$$

Since Z is arbitrary, we conclude $a^j(\vec{x}, \vec{x}') = X^j(\vec{x})$ and $b_{j'}(\vec{x}, \vec{x}') = -x_{i'}\partial_{x_j}X^i$ so that

$$X^C = X^j(\vec{x})\partial_{x_j} - x_{i'}(\partial_{x_j}X^i)(\vec{x})\partial_{x_{j'}} . \tag{1.1.g}$$

Taking $X = \partial_{x_j}$ then yields $X^C = \partial_{x_j}$. Since $\omega \neq 0$, we have $x_{i'} \neq 0$ for some i. Taking $X = x^j\partial_{x_i}$ then yields $X^C = x^j\partial_{x_i} - x_{i'}\partial_{x_{j'}}$ which completes the proof. We note that this fails on the 0-section; if $\omega = 0$, then

$$\text{Span}\{X^C\} = \text{Span}\{\partial_{x_j}\}. \qquad\square$$

Remark 1.3   A crucial point is that, since ιX and X^C are invariantly characterized, we do not need to check the local formalism transforms correctly; on the other hand, the local formalism shows that X^C in fact exists. We shall apply similar arguments subsequently.

We use Lemma 1.2 to prove the following result—see also Yano and Ishihara [215]:

Lemma 1.4   *Two smooth tensor fields Ψ_1, Ψ_2 of type $(0, s)$ on T^*M coincide with each other if and only if we have the following identity for all vector fields X_i on M:*

$$\Psi_1(X^C_{i_1}, \ldots, X^C_{i_s}) = \Psi_2(X^C_{i_1}, \ldots, X^C_{i_s}).$$

Proof. This is immediate if $\omega \neq 0$ by Lemma 1.2; we now use continuity to extend the result to the 0-section.   □

We use Lemma 1.4 to show that the symplectic 2-form $\Omega = dx_{i'} \wedge dx^i$ of Equation (1.1.f) is independent of the system of local coordinates. By Equation (1.1.g),

$$
\begin{aligned}
\Omega(X^C, Y^C) &= \Omega(X^j \partial_{x_j} - x_{i'}(\partial_{x_j} X^i)\partial_{x_{j'}}, Y^j \partial_{x_j} - x_{i'}(\partial_{x_j} Y^i)\partial_{x_{j'}}) \\
&= x_{i'}\{-Y^j \partial_{x_j} X^i + X^j \partial_{x_j} Y^i\}, \\
\iota[X, Y] &= \iota((X^j \partial_{x_j} Y^i - Y^j \partial_{x_j} X^i)\partial_{x_i}) \\
&= x_{i'}\{-Y^j \partial_{x_j} X^i + X^j \partial_{x_j} Y^i\}.
\end{aligned}
$$

Consequently, the symplectic 2-form Ω of Equation (1.1.f) may be characterized invariantly by the identity:

$$\Omega(X^C, Y^C) = \iota[X, Y]. \tag{1.1.h}$$

Let $T \in C^\infty(\mathrm{End}\{TM\})$ be a smooth $(1, 1)$-tensor field on M. Define a corresponding lifted 1-form $\iota T \in C^\infty(T^*(T^*M))$ on the cotangent bundle that is characterized by the identity:

$$\ll \iota T, X^C \gg = \iota(TX).$$

We compute in local coordinates as follows. Expand $\iota T = a_i(\vec{x}, \vec{x}')dx^i + b^i(\vec{x}, \vec{x}')dx_{i'}$ and $T\partial_{x_i} = T_i^j \partial_{x_j}$. Let $X = X^i \partial_{x_i}$. We compute:

$$
\begin{aligned}
(\iota T)(X^C) &= a_i(\vec{x}, \vec{x}')X^i(\vec{x}) - b^j(\vec{x}, \vec{x}')x_{i'}(\partial_{x_j} X^i), \\
\iota(TX) &= \iota(X^i T_i^j \partial_{x_j}) = x_{j'}X^i T_i^j.
\end{aligned}
$$

This shows that $b^j = 0$ and that $a_i = x_{j'}T_i^j$, i.e., that

$$\iota T = x_{j'}T_i^j dx^i \text{ where } T = T_i^j \partial_{x_j} \otimes dx^i.$$

We also refer to Kowalski and Sekizawa [145, 146] for additional information concerning natural lifts in Riemannian geometry. The following result summarizes the formalisms that we have established in this section:

Lemma 1.5

1. *The evaluation map ιX for $X \in C^\infty(TM)$:*

    (a) *Invariant formalism:* $\iota X(P, \omega) = \omega(X_P)$.

    (b) *Coordinate formalism:* $\iota X = x_{i'}X^i$ *for* $X = X^i \partial_{x_i}$.

2. *The complete lift of a vector field $X \in C^\infty(TM)$.*

   (a) *Invariant formalism: $X^C(\iota Z) = \iota[X, Z]$ for $Z \in C^\infty(TM)$.*

   (b) *Coordinate formalism: $X^C = X^j \partial_{x_j} - x_{i'}(\partial_{x_j} X^i)\partial_{x_{j'}}$ for $X = X^j \partial_{x_j}$.*

3. *Symplectic form Ω on T^*M.*

   (a) *Invariant formalism: $\Omega(X^C, Y^C) = \iota[X, Y]$.*

   (b) *Coordinate formalism: $\Omega = dx_{i'} \wedge dx^i$.*

4. *Let $T \in C^\infty(\text{End}\{TM\})$; ιT is the 1-form on T^*M:*

   (a) *Invariant formalism: $\ll \iota T, X^C \gg = \iota(TX)$.*

   (b) *Coordinate formalism: $\iota T = x_{j'}T_i^j dx^i$.*

1.2   CONNECTIONS

The study of an invariantly defined directional derivative probably began with Christoffel [54]. Subsequently, additional work was done by Ricci and Levi-Civita [186, 188]. We also refer to early work by Cartan [48]. If $\mathbb{V}$ is a vector bundle over M, let $C^\infty(\mathbb{V})$ denote the space of smooth sections to $\mathbb{V}$. A connection ∇ on $\mathbb{V}$ is a first-order partial differential operator from $C^\infty(\mathbb{V})$ to $C^\infty(T^*M \otimes \mathbb{V})$, which satisfies the *Leibnitz formula*:

$$\nabla(fs) = df \otimes s + f\nabla s \quad \text{for} \quad s \in C^\infty(\mathbb{V}). \tag{1.2.a}$$

The associated *directional covariant derivative* $\nabla_X s$ is defined by setting

$$\nabla_X s := \ll X, \nabla s \gg \text{ for } s \in C^\infty(\mathbb{V}) \text{ and } X \in C^\infty(TM),$$

where we use $\ll \cdot, \cdot \gg$ to denote the natural pairing from $TM \otimes T^*M \otimes \mathbb{V}$ to $\mathbb{V}$. If $\{e_i\}$ is a basis for TM and if $\{e^i\}$ is the associated *dual basis* for T^*M, the total covariant derivative is then given in terms of the directional covariant derivatives by

$$\nabla s = e^i \otimes \nabla_{e_i} s,$$

where, as noted previously, we adopt the *Einstein convention* and sum over repeated indices. If $\mathbb{V}$ is a complex vector bundle, we require ∇ to be complex linear. We extend ∇ to a *dual connection* ∇^* on $\mathbb{V}^*$ by requiring that

$$d \ll s, s^* \gg = \ll \nabla s, s^* \gg + \ll s, \nabla^* s^* \gg \text{ for all } s \in C^\infty(\mathbb{V}) \text{ and } s^* \in C^\infty(\mathbb{V}^*).$$

Let $\vec{s} = (s_1, \ldots, s_k)$ be a local frame for $\mathbb{V}$ and let $\vec{x} = (x^1, \ldots, x^m)$ be a system of local coordinates on M. We may expand

$$\nabla_{\partial_{x_i}} s_a = \Gamma_{ia}{}^b s_b \,.$$

Here, the index i ranges from 1 to $m := \dim(M)$ and the indices a, b range from 1 to k, i.e., to the fiber dimension of $\mathbb{V}$. The $\Gamma_{ia}{}^b = {}^\nabla\Gamma_{ia}{}^b$ are referred to as the *Christoffel symbols of the first kind* or sometimes simply as the *Christoffel symbols* of the connection ∇. They are not tensorial. In view of the *Leibnitz formula* given in Equation (1.2.a), ∇ is determined by the Christoffel symbols as we see by computing:

$$\nabla_{\alpha^i \partial_{x_i}} (v^a s_a) = \alpha^i \{ v^a \Gamma_{ia}{}^b s_b + \partial_{x_i}(v^a) s_a \} \,.$$

Let $\vec{s}\,^* = (s^1, \ldots, s^k)$ be the local *dual frame* field for the dual bundle $\mathbb{V}^*$. The *dual Christoffel symbols* for the *dual connection* on $\mathbb{V}^*$ are given by the identity:

$$\nabla^*_{\partial_{x_i}} s^b = -\Gamma_{ia}{}^b s^a \,.$$

If $\mathbb{V}$ is equipped with a non-degenerate inner product or *fiber metric h*, then we say that ∇ is a *Riemannian connection* in the real setting or a *Hermitian connection* in the complex setting if the following relation holds:

$$dh(s_1, s_2) = h(\nabla s_1, s_2) + h(s_1, \nabla s_2) \text{ for } s_i \in C^\infty(\mathbb{V}) \,.$$

Equivalently, ∇ is Riemannian if and only if ∇ agrees with ∇^* when we use h to identify $\mathbb{V}$ with $\mathbb{V}^*$. We can lower indices and define the *Christoffel symbols of the second kind* by setting:

$$\Gamma_{iab} := h(\nabla_{\partial_{x_i}} s_a, s_b) \,.$$

If $\vec{s}$ is a local orthonormal frame for $\mathbb{V}$, then ∇ is Riemannian if and only if we have the symmetry:

$$\Gamma_{iab} + \Gamma_{iba} = 0 \text{ for all } i, a, b \,.$$

THE CURVATURE OPERATOR

We refer to Besse [12], to Kobayashi and Nomizu [140], to Riemann [189], and to Ricci and Levi-Civita [188] for further details concerning the material of this section. The *curvature operator* $\mathcal{R} = {}^\nabla\mathcal{R}$ of a connection ∇ is defined by setting:

$$\mathcal{R}(X, Y)s := \{ \nabla_X \nabla_Y - \nabla_Y \nabla_X - \nabla_{[X,Y]} \} s \,. \tag{1.2.b}$$

This is a *tensor*, i.e., if $X, Y \in C^\infty(TM)$, $s \in C^\infty(\mathbb{V})$, and $f \in C^\infty(M)$, then:

$$\mathcal{R}(fX, Y)s = \mathcal{R}(X, fY)s = \mathcal{R}(X, Y)fs = f\mathcal{R}(X, Y)s \,.$$

Let $\mathcal{R}(\partial_{x_i}, \partial_{x_j})s_a = R_{ija}{}^b s_b$ be the *components of the curvature operator* in a system of local coordinates $\vec{x} = (x^1, \ldots, x^m)$ and relative to a local frame $\vec{s} = (s_1, \ldots, s_k)$ for $\mathbb{V}$. We have

$$R_{ija}{}^b = \partial_{x_i} \Gamma_{ja}{}^b - \partial_{x_j} \Gamma_{ia}{}^b + \Gamma_{ic}{}^b \Gamma_{ja}{}^c - \Gamma_{jc}{}^b \Gamma_{ia}{}^c \,. \tag{1.2.c}$$

If h is a non-degenerate inner product on $\mathbb{V}$, we may lower indices to define:

$$R(X, Y, s, \tilde{s}) := h(\mathcal{R}(X, Y)s, \tilde{s}) \text{ and set } R_{ijab} := h(\mathcal{R}(\partial_{x_i}, \partial_{x_j})s_a, s_b). \qquad (1.2.d)$$

If ∇ is Riemannian, and if $\vec{s}$ is a local orthonormal frame for $\mathbb{V}$, then:

$$R_{ijab} + R_{ijba} = 0.$$

AFFINE CONNECTIONS

Of particular interest is the special case in which $\mathbb{V}$ is the tangent bundle of M and we shall restrict to this case henceforth unless otherwise noted. Let ∇ be a connection on TM. Define the *Ricci tensor* $\rho = {}^\nabla\rho$, the *symmetric Ricci tensor* $\rho_s = {}^\nabla\rho_s$, and the *alternating Ricci tensor* $\rho_a = {}^\nabla\rho_a$ by setting:

$$\begin{aligned}
\rho(X, Y) &:= \operatorname{Tr}\{Z \to \mathcal{R}(Z, X)Y\}, \\
\rho_s(X, Y) &:= \tfrac{1}{2}\{\rho(X, Y) + \rho(Y, X)\}, \\
\rho_a(X, Y) &:= \tfrac{1}{2}\{\rho(X, Y) - \rho(Y, X)\}.
\end{aligned}$$

Let $\mathcal{M} = (M, g)$ be a *pseudo-Riemannian manifold*; here $g \in C^\infty(S^2(T^*M))$ is a non-degenerate symmetric bilinear form—see Eisenhart [77, 78] for details. We say that $\mathcal{M}$ is *Riemannian* if g is positive definite and that $\mathcal{M}$ is *Lorentzian* if g has signature $(1, m-1)$. We can raise indices to define the *Ricci operator* $\operatorname{Ric} = {}^\nabla\operatorname{Ric}$. The Ricci operator is a smooth section to $C^\infty(\operatorname{End}\{TM\})$, which is characterized by the identity:

$$g(\operatorname{Ric} X, Y) = \rho(X, Y).$$

The Ricci tensor of the Levi-Civita connection is always a symmetric 2-cotensor; a pseudo-Riemannian manifold is said to be *Einstein* if the Ricci tensor is a constant multiple of the metric tensor; we refer to [12, 78, 140, 187] for further details.

If ∇ is a connection on TM, then we may define the *torsion tensor* $\mathcal{T} = {}^\nabla\mathcal{T}$ by:

$$\mathcal{T}(X, Y) = \nabla_X Y - \nabla_Y X - [X, Y] \text{ for } X, Y \in C^\infty(TM).$$

The torsion is a tensor, i.e.,

$$\mathcal{T}(fX, Y) = \mathcal{T}(X, fY) = f\mathcal{T}(X, Y) \text{ for } X, Y \in C^\infty(TM) \text{ and } f \in C^\infty(M).$$

One has the following useful fact; it permits one to normalize the choice of the frame so that only the second derivatives of the Christoffel symbols enter into the computation of the curvature:

Lemma 1.6   *Let ∇ be a connection on TM and let $P \in M$. The following conditions are equivalent and if either condition is satisfied at all points of M, then ∇ is said to be an* affine connection *or, equivalently, to be a* torsion-free connection.

1. $\mathcal{T}(X_P, Y_P) = 0$ for all $X_P, Y_P \in T_P M$.

2. There exist local coordinates for M centered at P so that $^\nabla\Gamma(P) = 0$.

Proof. It is immediate that the torsion tensor $\mathcal{T}$ vanishes at P if and only if we have the symmetry $\Gamma_{ij}{}^k(P) = \Gamma_{ji}{}^k(P)$. In particular, if there exists a coordinate system where $\Gamma(P) = 0$, then necessarily $\mathcal{T}$ vanishes at P. Thus, Assertion (2) implies Assertion (1). Conversely, assume that Assertion (1) holds. Choose any system of coordinates $\vec{x} = (x^1, \ldots, x^m)$ on M which are centered at P. Define a new system of coordinates by setting:

$$z^i = x^i + \tfrac{1}{2} \sum_{j,k} c_{jk}{}^i \, x^j x^k \,,$$

where $c_{jk}{}^i = c_{kj}{}^i$ remains to be chosen. As $\partial_{x_j} = \partial_{z_j} + c_{ji}{}^\ell x^i \partial_{z_\ell}$,

$$\nabla_{\partial_{x_i}} \partial_{x_j}(P) = \nabla_{\partial_{z_i}} \partial_{z_j}(P) + c_{ji}{}^\ell \partial_{z_\ell}(P) \,.$$

Set $c_{ij}{}^\ell := \Gamma_{ij}{}^\ell(P)$; the fact that $c_{ij}{}^\ell = c_{ji}{}^\ell$ is exactly the assumption that the torsion tensor of ∇ vanishes at P. Thus we conclude that the new coordinate system has vanishing Christoffel symbols at P. □

Although there are interesting examples where ∇ has torsion (see, for example, Cartan [47] and Ivanov [133]), we shall henceforth assume ∇ is torsion-free. We say that (M, ∇) is an *affine manifold* if ∇ is an affine connection. See, for example, the discussion in [14, 129, 168, 190, 192, 193]. In this setting, the curvature operator has the following symmetries:

$$\begin{aligned} \mathcal{R}(X, Y)Z + \mathcal{R}(Y, X)Z &= 0, & (1.2.\text{e}) \\ \mathcal{R}(X, Y)Z + \mathcal{R}(Y, Z)X + \mathcal{R}(Z, X)Y &= 0\,. & (1.2.\text{f}) \end{aligned}$$

The symmetry of Equation (1.2.f) is called the *first Bianchi identity* or sometimes just the *Bianchi identity*—see Bianchi [13]. The *first covariant derivative* of the curvature operator is defined by setting:

$$\begin{aligned} \nabla\mathcal{R}(X_1, X_2; X_4)X_3 &:= \nabla_{X_4}\mathcal{R}(X_1, X_2)X_3 - \mathcal{R}(\nabla_{X_4}X_1, X_2)X_3 \\ &\quad - \mathcal{R}(X_1, \nabla_{X_4}X_2)X_3 - \mathcal{R}(X_1, X_2)\nabla_{X_4}X_3 \,. \end{aligned} \qquad (1.2.\text{g})$$

$\nabla\mathcal{R}$ is a tensor which belongs to $\otimes^3 T^*M \otimes \operatorname{End}\{TM\}$ and which has the symmetries:

$$\begin{aligned} \nabla\mathcal{R}(X_1, X_2; X_4)X_3 + \nabla\mathcal{R}(X_2, X_1; X_4)X_3 &= 0, & (1.2.\text{h}) \\ \nabla\mathcal{R}(X_1, X_2; X_4)X_3 + \nabla\mathcal{R}(X_2, X_3; X_4)X_1 + \nabla\mathcal{R}(X_3, X_1; X_4)X_2 &= 0, & (1.2.\text{i}) \\ \nabla\mathcal{R}(X_1, X_2; X_4)X_3 + \nabla\mathcal{R}(X_2, X_4; X_1)X_3 + \nabla\mathcal{R}(X_4, X_1; X_2)X_3 &= 0\,. & (1.2.\text{j}) \end{aligned}$$

Equation (1.2.i) arises as the *covariant derivative of the first Bianchi identity*; Equation (1.2.j) is known as the *second Bianchi identity*. We can continue this process to define higher covariant derivatives $\nabla^k\mathcal{R} \in \otimes^{2+k}(T^*M) \otimes \operatorname{End}\{TM\}$ for any k.

We say (M, ∇) is *flat* if $\mathcal{R} = 0$ and that (M, ∇) is *locally symmetric* if $\nabla \mathcal{R} = 0$. The following result of Schirokow [190] is a useful observation:

Theorem 1.7   *Let ∇ be an affine connection on TM. The following assertions are equivalent. If any is satisfied, then ∇ is said to be equiaffine or Ricci symmetric.*

1. *Let $\omega := \Gamma_{ij}{}^j dx^i$. Then $d\omega_{\tilde{x}} = 0$ for any system of local coordinates $\tilde{x}$ on M.*

2. $\mathrm{Tr}\{\mathcal{R}\} = 0$.

3. *The connection ∇ is Ricci symmetric.*

4. *The connection ∇ locally admits a parallel volume form.*

Proof. We use the Bianchi identity of Equation (1.2.f) to see:

$$
\begin{aligned}
0 &= \mathrm{Tr}\{Z \to \mathcal{R}(X, Y)Z\} + \mathrm{Tr}\{Z \to \mathcal{R}(Y, Z)X\} + \mathrm{Tr}\{Z \to \mathcal{R}(Z, X)Y\} \\
&= \mathrm{Tr}\{\mathcal{R}(X, Y)\} - \rho(Y, X) + \rho(X, Y) = 0 \,.
\end{aligned}
$$

This shows that Assertion (2) and Assertion (3) are equivalent. We show that Assertion (1) and Assertion (2) are equivalent by computing:

$$
\begin{aligned}
\mathrm{Tr}\{\mathcal{R}_{ij}\} &dx^i \wedge dx^j \\
&= \{\partial_{x_i} \Gamma_{jk}{}^k - \partial_{x_j} \Gamma_{ik}{}^k + \Gamma_{in}{}^k \Gamma_{jk}{}^n - \Gamma_{jn}{}^k \Gamma_{ik}{}^n\} dx^i \wedge dx^j \\
&= \{\partial_{x_i} \Gamma_{jk}{}^k - \partial_{x_j} \Gamma_{ik}{}^k\} dx^i \wedge dx^j = 2d\{\Gamma_{jk}{}^k dx^j\} \,.
\end{aligned}
$$

Let e^Φ be a conformal factor to rescale the volume form. Since $\nabla^*_{\partial_{x_i}} dx^j = -\Gamma_{ik}{}^j dx^k$, we may compute:

$$
\nabla^*_{\partial_{x_i}} \{e^\Phi dx^1 \wedge \cdots \wedge dx^m\} = \{\partial_{x_i} \Phi - \textstyle\sum_k \Gamma_{ik}{}^k\}\{e^\Phi dx^1 \wedge \cdots \wedge dx^m\} \,.
$$

Thus, there exists a local parallel volume form on an open subset of M if and only if $\Gamma_{ik}{}^k dx^i$ is exact on that open subset. As every closed 1-form is locally exact, Assertion (1) and Assertion (4) are equivalent. □

Geodesics

We say that a parametrized curve $\gamma(t)$ in M is a *geodesic* if it satisfies the *geodesic equation*

$$
\nabla_{\dot\gamma} \dot\gamma = 0 \,,
$$

or, equivalently, in a system of coordinates where $\gamma = (\gamma^1, \ldots, \gamma^m)$ we have:

$$
\ddot\gamma^i + \Gamma_{jk}{}^i \dot\gamma^j \dot\gamma^k = 0 \,.
$$

In the case of the Levi-Civita connection for a positive definite metric, which will be discussed presently, this means that γ locally minimizes distance between points on γ. An affine connection ∇ is said to be *geodesically complete* if every geodesic extends to the parameter range $(-\infty, \infty)$.

The *exponential map* $\exp_P : T_P M \to M$ is a local diffeomorphism from a neighborhood of the origin $0 \in T_P M$ to a neighborhood of P in M. It is characterized by the fact that the curves $\gamma_v(s) := \exp_P(sv)$ are geodesics in M with initial velocity $v \in T_P M$. One says that (M, ∇) is *complete* if $\exp_P$ is defined on all $T_P M$. *Conjugate points* arise where $\exp_P$ fails to be a local diffeomorphism. Furthermore, since there can be different geodesics joining two points, $\exp_P$ can fail to be globally one-to-one.

Projective Equivalence and the Projective Curvature Tensor

We refer to Nomizu [166] for further details concerning the following material.

Lemma 1.8   *Let ∇ and $\tilde{\nabla}$ be connections on TM. Let $S(X, Y) = \nabla_X Y - \tilde{\nabla}_X Y$ be the difference tensor.*

1. *The following conditions are equivalent and if any is satisfied, then the two connections are said to be* projectively equivalent:

    (a) *∇ and $\tilde{\nabla}$ have the same unparametrized geodesics.*

    (b) *There exists a smooth 1-form ω such that $S(X, X) = 2\omega(X)X$ for all X.*

2. *If ∇ and $\tilde{\nabla}$ are torsion-free, then ∇ and $\tilde{\nabla}$ are projectively equivalent if and only if $S(X, Y) = \omega(X)Y + \omega(Y)X$.*

3. *If $\tilde{\nabla}$ is an arbitrary connection with torsion tensor $\tilde{T}$, then setting $S = -\frac{1}{2}\tilde{T}$ yields a torsion-free connection ∇ which is projectively equivalent to $\tilde{\nabla}$.*

Proof. Let $\gamma(t)$ be a geodesic for ∇ and let $\mu(s) = \gamma(\alpha(s))$ be a reparametrization of γ. Then $\dot{\mu}(s) = \dot{\alpha}(s)\dot{\gamma}(\alpha(s))$. Since $\nabla_{\dot\gamma}\dot\gamma = 0$,

$$\tilde{\nabla}_{\dot\mu}\dot\mu = (\dot{\alpha})^2 S(\dot\gamma, \dot\gamma) + \dot{\alpha}\ddot{\alpha}\dot\gamma \,.$$

Thus, it is possible to reparametrize γ to obtain a geodesic for $\tilde{\nabla}$ if and only if $S(\dot\gamma, \dot\gamma)$ is a multiple of $\dot\gamma$; Assertion (1) now follows. If ∇ and $\tilde{\nabla}$ are torsion-free, then the tensor S is symmetric; Assertion (2) now follows by polarization. Set $S = -\frac{1}{2}\tilde{T}$. Because $\tilde{T}$ is a skew-symmetric tensor, $S(X, X) = 0$, so ∇ is projectively equivalent to $\tilde{\nabla}$. Assertion (3) follows as the torsion tensor for ∇ is given by $\tilde{T} + 2S = 0$. □

If the Ricci tensor $^\nabla\rho$ is a symmetric tensor, then the *Weyl projective curvature operator* is defined by setting:

$$\mathcal{P}(X, Y)Z := \mathcal{R}(X, Y)Z - \frac{1}{m-1}\{\rho(Y, Z)X - \rho(X, Z)Y\} \,.$$

We note that if ∇ and $\tilde{\nabla}$ are projectively equivalent, then the corresponding Weyl projective curvature operators coincide. We may show that the Weyl projective curvature operator vanishes if $m = 2$ by noting that the curvature operator of any affine surface (Σ, ∇) with symmetric Ricci tensor satisfies

$$\mathcal{R}(X, Y)Z = \rho(Y, Z)X - \rho(X, Z)Y.$$

One says that ∇ is *projectively flat* if around each point there is a projective change of ∇ to a flat affine connection. One has (see Eisenhart [77, Section 32]):

Lemma 1.9   *Let (M, ∇) be an affine manifold with symmetric Ricci tensor. Then ∇ is projectively flat if and only if one of the following holds*

1. $\dim(M) \geq 3$ *and* $\mathcal{P} = 0$.

2. $\dim(M) = 2$ *and* $(\nabla_X \rho)(Y, X) = (\nabla_Y \rho)(X, Z)$ *for all X, Y, Z vector fields on M.*

We note that the Ricci tensor of any projectively flat affine manifold (M, ∇) satisfies the identity in Assertion (2) above.

THE LEVI-CIVITA CONNECTION

Let $\mathcal{M} := (M, g)$ be a pseudo-Riemannian manifold. The *Levi-Civita connection* $\nabla = {}^g\nabla$ is the unique affine connection which is Riemannian; it is characterized by the identities:

$$Xg(Y, Z) = g(\nabla_X Y, Z) + g(Y, \nabla_X Z) \quad \text{and} \quad \nabla_X Y - \nabla_Y X = [X, Y],$$

for all $X, Y, Z \in C^\infty(TM)$. We refer to [7, 54, 140, 151, 196] for further details.

Let $\vec{x} = (x^1, \ldots, x^m)$ be a system of local coordinates on M, and let $g_{ij} := g(\partial_{x_i}, \partial_{x_j})$. We extend g dually to a metric on the cotangent bundle; $g^{ij} := g(dx^i, dx^j)$ is the inverse matrix. Let $\Gamma_{ij}{}^k = {}^g\Gamma_{ij}{}^k$ and $\Gamma_{ijk} = {}^g\Gamma_{ijk}$ be the associated Christoffel symbols of ${}^g\nabla$. We have the *Koszul formula*:

$$\Gamma_{ijk} = \tfrac{1}{2}\{\partial_{x_i} g_{jk} + \partial_{x_j} g_{ik} - \partial_{x_k} g_{ij}\} \quad \text{and} \quad \Gamma_{ij}{}^k = g^{k\ell}\Gamma_{ij\ell}. \tag{1.2.k}$$

Similarly, if $\{e_i\}$ is a local orthonormal frame for TM, let $g([e_i, e_j], e_k) = C_{ijk}$ give the *structure constants* for the *Lie bracket*. We have

$$\Gamma_{ijk} = \tfrac{1}{2}\{C_{ijk} - C_{jki} + C_{kij}\}.$$

The *curvature tensor R* of the Levi-Civita connection has the symmetries:

$$R(X, Y, Z, W) + R(Y, X, Z, W) = 0, \tag{1.2.l}$$
$$R(X, Y, Z, W) + R(Y, Z, X, W) + R(Z, X, Y, W) = 0, \tag{1.2.m}$$
$$R(X, Y, Z, W) + R(X, Y, W, Z) = 0, \tag{1.2.n}$$
$$R(X, Y, Z, W) - R(Z, W, X, Y) = 0. \tag{1.2.o}$$

The symmetries of Equation (1.2.l) and Equation (1.2.m) arise from the symmetries of Equation (1.2.e) and Equation (1.2.f) since ∇ is torsion-free. The symmetry of Equation (1.2.n) arises from the fact that ∇ is Riemannian. Note that the symmetry of Equation (1.2.n) and the symmetry of Equation (1.2.o) are equivalent in the presence of Equation (1.2.l) and Equation (1.2.m)—see, for example, the discussion in Blažić *et al.* [20]. If $\mathbb{V}$ is a vector bundle over M, let $S^2 = S^2(\mathbb{V}^*)$ (resp. $\Lambda^2 = \Lambda^2(\mathbb{V}^*)$) be the bundle of symmetric (resp. alternating) bilinear forms on $\mathbb{V}$. Let $S^2 M = S^2(T^*M)$ and $\Lambda^2 M = \Lambda^2(T^*M)$. The symmetries of Equation (1.2.l), Equation (1.2.o), and Equation (1.2.n) permit us to regard

$$R \in C^\infty(S^2(\Lambda^2 M)) . \tag{1.2.p}$$

Let $\{x, y\}$ be a basis for a non-degenerate 2-plane $\pi \subset T_P M$. The *sectional curvature* $\kappa(\pi)$ is defined by setting

$$\kappa(\pi) := \frac{R(x, y, y, x)}{g(x, x)g(y, y) - g(x, y)^2} .$$

The sectional curvatures determine the complete curvature tensor. One has (see, for example, Kobayashi and Nomizu [140] and O'Neill [169]):

Lemma 1.10 *Suppose that R and $\tilde{R}$ satisfy the identities of Equations (1.2.l)–(1.2.o) above. If $\kappa_R(\pi) = \kappa_{\tilde{R}}(\pi)$ for all non-degenerate 2-planes π, then $R = \tilde{R}$.*

We shall discuss the geometry of pseudo-Riemannian manifolds with *constant sectional curvature* subsequently in Section 1.10—see Wolf [212] for further details. A pseudo-Riemannian manifold is said to be *locally symmetric* if $\nabla R = 0$ (see, for example, Cartan [49, 50] and Helgason [123]); we will discuss the geometry of such manifolds as well, subsequently in Section 1.10.

WEYL GEOMETRY

Weyl geometry (see Weyl [209]) is, in a sense, midway between Riemannian geometry and affine geometry. The literature is a vast one and we can cite only a few references [23, 24, 46, 59, 132, 179, 180, 181, 199] in the geometric setting and [108, 109] in the algebraic setting.

Let (M, g) be a pseudo-Riemannian manifold. A tensor field K is said to be *recurrent* if there exists a 1-form ω such that

$$\nabla K = \omega \otimes K, \text{ i.e.}, \nabla_X K = \omega(X)K \text{ for all } X \in C^\infty(TM) .$$

An affine connection ∇ on TM defines a *Weyl structure* and (M, g, ∇) is said to be a *Weyl manifold* if the metric is recurrent, i.e., if there exists a smooth 1-form ϕ on M so that the structures are related by the equation:

$$\nabla g = -2\phi \otimes g .$$

If $^g\nabla$ is the Levi-Civita connection, then we may express $\nabla = {}^\phi\nabla$ in the form:

$$^\phi\nabla_X Y = {}^g\nabla_X Y + \phi(X)Y + \phi(Y)X - g(X, Y)\phi^\# , \tag{1.2.q}$$

where $\phi^{\#}$ is the dual vector field. Conversely, if ϕ is given and if we use Equation (1.2.q) to define ∇, then ∇ is a *Weyl connection* with associated 1-form ϕ. We refer to Gilkey, Nikčević, and Simon [109], to Ovando [174], and to Özdeǧer [175] for further details concerning Weyl geometry.

There is an additional curvature symmetry which pertains in Weyl geometry (see, for example, the discussion in Gilkey, Nikčević, and Simon [109]):

$$R(X, Y, Z, W) + R(X, Y, W, Z) = -\tfrac{4}{m}\rho_a(X, Y)g(Z, W) \,, \qquad (1.2.r)$$

where $\dim(M) = m$. The defining 1-form ϕ is related to the curvature by the equation:

$$d\phi = -\tfrac{2}{m}\,\rho_a \,. \qquad (1.2.s)$$

Weyl geometry is a conformal theory; if $g_1 = e^{2f}g$ is *conformally equivalent* to g and if (M, g, ∇) is a Weyl manifold, then (M, g_1, ∇) is again a Weyl manifold with associated 1-form ϕ_1 given by $\phi_1 = \phi - df$. One has the following well known result characterizing *trivial Weyl structures* (see, for example, Gilkey, Nikčević, and Simon [109]):

Theorem 1.11 *Let (M, g, ∇) be a Weyl manifold with $H^1_{\mathrm{DeR}}(M) = 0$. The following assertions are equivalent and if any is satisfied, then the Weyl structure is said to be trivial.*

1. *$d\phi = 0$.*

2. *$\nabla = {}^{g_1}\nabla$ for some conformally equivalent pseudo-Riemannian metric g_1.*

3. *$\nabla = {}^{g_1}\nabla$ for some pseudo-Riemannian metric g_1.*

4. *ρ is a symmetric tensor.*

Proof. Suppose that $d\phi = 0$. Since the *first de Rham cohomology group* of M vanishes by assumption, there exists $f \in C^\infty(M)$ so that $\phi = df$. We define the *conformally equivalent metric* $g_1 := e^{2f}g$. It is then clear that $\nabla g_1 = e^{2f}(2df - 2\phi) = 0$. Thus, ∇ is torsion-free and Riemannian with respect to g_1; this implies $\nabla = {}^{g_1}\nabla$ is the Levi-Civita connection defined by g_1. Thus, Assertion (1) implies Assertion (2); Assertion (2) trivially implies Assertion (3). Since the curvature tensor of the Levi-Civita connection has a symmetric Ricci tensor, Assertion (3) implies Assertion (4). Finally, suppose that Assertion (4) holds so that $\rho_a = 0$. By Equation (1.2.s), $d\phi = -\tfrac{2}{m}\rho_a = 0$. $\square$

1.3 CURVATURE MODELS IN THE REAL SETTING

Let $(V, \langle \cdot, \cdot \rangle)$ be an *inner product space*. Here V is a real vector space of dimension m and $\langle \cdot, \cdot \rangle$ is a non-degenerate, symmetric, bilinear form on V of signature (p, q) where $m = p + q$. A 4-tensor $A \in \otimes^4 V^*$ is said to be a *Riemannian algebraic curvature tensor* if A satisfies the symmetries given in Equation (1.2.l), in Equation (1.2.m), and in Equation (1.2.n); note that Equation (1.2.o) is then satisfied automatically. Let $\mathfrak{R}(V)$ be the subspace of $\otimes^4 V^*$, which consists of all tensors satisfying these relations. We say that a triple $(V, \langle \cdot, \cdot \rangle, A)$ is a *Riemannian curvature model* if $A \in \mathfrak{R}(V)$. One says that such a triple is *geometrically realizable* by a pseudo-Riemannian manifold if there is a point P of some pseudo-Riemannian manifold (M, g) and if there is an isomorphism $\Phi : V \to T_P M$ so:

$$\Phi^* g_P = \langle \cdot, \cdot \rangle \quad \text{and} \quad \Phi^* {}^g R_P = A \,.$$

If $A \in \text{End}\{V\} \otimes V^*$ only satisfies the identities of Equation (1.2.e) and of Equation (1.2.f), then A will be said to be an *affine curvature tensor* and the pair (V, A) will be said to be an *affine curvature model*; see Blažić *et al.* [20] and Gilkey, Nikčević, and Simon [107] for further details. Such a pair is said to be *geometrically realizable* by an affine manifold if there is a point P of some affine manifold (M, ∇) and if there is an isomorphism $\Phi : V \to T_P M$ so that $\Phi^* {}^\nabla R_P = A$. In the presence of a metric, we can lower indices and let $\mathfrak{A}(V) \subset \otimes^4 V^*$ be the space of all tensors satisfying Equation (1.2.l) and Equation (1.2.m).

Finally, let $\mathfrak{W}(V) \subset \otimes^4 V^*$ be the space of 4-tensors satisfying Equation (1.2.l), Equation (1.2.m), and Equation (1.2.r); these are the *Weyl algebraic curvature tensors*—see, for example, Nomizu [167]. If $A \in \mathfrak{R}(V)$, then $\rho_a = 0$ and $R(X, Y, Z, W) + R(X, Y, W, Z) = 0$. Consequently,

$$\mathfrak{R}(V) \subset \mathfrak{W}(V) \subset \mathfrak{A}(V) \,.$$

A triple $(V, \langle \cdot, \cdot \rangle, A)$ is said to be a *Weyl curvature model* if $A \in \mathfrak{W}(V)$. The notion of geometric realizability is defined analogously in this setting. We refer to Blažić *et al.* [20], to Gilkey [98], and to Gilkey, Nikčević, and Simon [109] for the proof of the following result; the first two assertions are, of course, well known:

Theorem 1.12

1. *Every Riemannian curvature model can be realized geometrically by a pseudo-Riemannian manifold.*

2. *Every affine curvature model can be realized geometrically by an affine manifold.*

3. *Every Weyl curvature model can be realized geometrically by a Weyl manifold.*

1.4   KÄHLER GEOMETRY

We now pass from the real to the complex setting and to the para-complex setting; we refer to Cruceanu, Fortuny, and Gadea [60] for details on para-complex geometry. Let V be a real vector space of even dimension $m = 2\bar{m}$. A *complex structure* on V is an endomorphism J_- of V so $J_-^2 = -\text{Id}$. Similarly, a *para-complex structure* on V is an endomorphism J_+ of V so $J_+^2 = \text{Id}$ and $\text{Tr}\{J_+\} = 0$; this trace-free condition is automatic in the complex setting, but must be imposed in the para-complex setting. It is convenient to introduce the notation $J_\pm$ in order to have a common formulation that permits us to treat both contexts in a similar fashion, although, we shall never be considering both structures simultaneously. In this setting, we say that $(V, J_\pm)$ is a *(para)-complex vector space*. In the geometric setting, $(M, J_\pm)$ is said to be an *almost (para)-complex manifold* if $J_\pm$ is a smooth endomorphism of the tangent bundle so that $(T_P M, J_\pm)$ is a (para)-complex structure for every $P \in M$. Define the *Nijenhuis tensor* $N_{J_\pm}$ by setting:

$$N_{J_\pm}(X, Y) := [X, Y] \mp J_\pm[J_\pm X, Y] \mp J_\pm[X, J_\pm Y] \pm [J_\pm X, J_\pm Y].$$

Theorem 1.13   *Let $(M, J_\pm)$ be a $2\bar{m}$-dimensional almost (para)-complex manifold. The following conditions are equivalent and if any is satisfied, then $(M, J_\pm)$ is said to be a (para)-complex manifold and the structure $J_\pm$ is said to be integrable:*

1. *There are coordinate charts $(x^1, y^1, \ldots, x^{\bar{m}}, y^{\bar{m}})$ covering M so that $J_\pm \partial_{x_i} = \partial_{y_i}$ and $J_\pm \partial_{y_i} = \pm \partial_{x_i}$ for $1 \leq i \leq \bar{m}$.*

2. *The Nijenhuis tensor $N_{J_\pm}$ vanishes.*

3. *The eigenbundles of $J_\pm$ are closed under the Lie bracket.*

The equivalence of these two definitions in the complex setting uses the *Newlander–Nirenberg theorem* [130, 157, 158, 206], which is a deep result in the theory of partial differential equations that can be regarded as a complex version of the Frobenius theorem. By contrast, the corresponding result in the para-complex setting only relies on the *Frobenius theorem* [57, 63, 81, 140, 147]—see, for example, the discussion in Brozos-Vázquez, Gilkey, and Nikčević [37] and in Cruceanu, Fortuny, and Gadea [60]. We refer to Frölicher and Nijenhuis [82, 83] and to Kobayashi and Nomizu [140] for further details concerning the Nijenhuis tensor.

Let $(M, J_\pm)$ be an almost (para)-complex manifold and let ∇ be a torsion-free connection on M. We use $J_\pm$ to give a (para)-complex structure to the tangent bundle. The triple $(M, J_\pm, \nabla)$ is said to be a *(para)-Kähler affine manifold* if ∇ is a (para)-complex connection; this means that $\nabla J_\pm = J_\pm \nabla$ or, equivalently, that the covariant derivative of $J_\pm$ vanishes. This condition then implies that $J_\pm$ is integrable and that

$$\mathcal{R}(X, Y)J_\pm = J_\pm \mathcal{R}(X, Y). \tag{1.4.a}$$

Working again in an algebraic setting, let $\langle \cdot, \cdot \rangle$ be a non-degenerate symmetric bilinear form of signature (p, q) on V. We say that the triple $(V, \langle \cdot, \cdot \rangle, J_\pm)$ is a *(para)-Hermitian vector space* if $J_\pm^* \langle \cdot, \cdot \rangle = \mp \langle \cdot, \cdot \rangle$. Thus, we permit the signature to be indefinite even when speaking of a Hermitian vector space. If J_+ defines a para-complex structure, necessarily $p = q$ so $\langle \cdot, \cdot \rangle$ has neutral signature. If J_- defines a complex structure, then necessarily p and q are both even.

Similarly, in the geometric setting, a triple $(M, g, J_\pm)$ is said to be an almost *(para)-Hermitian manifold* if $J_\pm$ is an almost (para)-complex structure, if (M, g) is a pseudo-Riemannian manifold, and if $J_{\pm}^* g = \mp g$. The triple is said to be a *(para)-Hermitian manifold* if $J_{\pm}$ is integrable. If $^g\nabla J_{\pm} = 0$, we can lower indices in Equation (1.4.a) to obtain the following symmetry, which is called the *(para)-Kähler identity*:

$$R(X, Y, Z, W) = \mp R(X, Y, J_{\pm}Z, J_{\pm}W).$$ (1.4.b)

A (para)-Hermitian pseudo-Riemannian manifold $(M, g, J_{\pm})$ will be said to be a *(para)-Kähler manifold* if $^g\nabla J_{\pm} = 0$ or, equivalently, if $d\Omega_{\pm} = 0$ where $\Omega_{\pm}$ is the *(para)-Kähler form*:

$$\Omega(X, Y) := g(X, J_{\pm}Y).$$

Thus, in particular, a (para)-Kähler manifold inherits a natural symplectic structure. A quadruple $(M, g, J_{\pm}, \nabla)$ is said to be a *(para)-Hermitian Weyl manifold* if (M, g, ∇) is a Weyl manifold and if $(M, g, J_{\pm})$ is a (para)-Hermitian manifold. The structure is said to be a *(para)-Kähler–Weyl* manifold if $\nabla J_{\pm} = 0$. We refer to [32, 102, 103, 105] for further details.

We now pass again to the algebraic context. Define the space of (para)-Kähler tensors $\mathfrak{K}_{\pm}$, the space of (para)-Kähler affine algebraic curvature tensors $\mathfrak{K}_{\pm,\mathfrak{A}}$, the space of (para)-Kähler Riemannian algebraic curvature tensors $\mathfrak{K}_{\pm,\mathfrak{R}}$, and the space of (para)-Kähler–Weyl algebraic curvature tensors $\mathfrak{K}_{\pm,\mathfrak{W}}$ by setting, respectively:

$$\mathfrak{K}_{\pm} := \{A \in \otimes^4 V^* : A(x, y, z, w) = \mp A(x, y, J_{\pm}z, J_{\pm}w)\},$$
$$\mathfrak{K}_{\pm,\mathfrak{A}} := \mathfrak{K}_{\pm} \cap \mathfrak{A}, \quad \mathfrak{K}_{\pm,\mathfrak{R}} := \mathfrak{K}_{\pm} \cap \mathfrak{R}, \quad \mathfrak{K}_{\pm,\mathfrak{W}} := \mathfrak{K}_{\pm} \cap \mathfrak{W}.$$

Let $(V, J_{\pm})$ be a (para)-complex vector space. The triple $(V, J_{\pm}, A)$ is said to be a *(para)-Kähler affine curvature model* if $A \in \mathfrak{K}_{\pm,\mathfrak{A}}$. Let $(V, \langle \cdot, \cdot \rangle, J_{\pm})$ be a (para)-Hermitian vector space. The quadruple $(V, \langle \cdot, \cdot \rangle, J_{\pm}, A)$ is said to be a *(para)-Kähler Hermitian curvature model* if $J_{\pm}^* \langle \cdot, \cdot \rangle = \mp \langle \cdot, \cdot \rangle$ and if $A \in \mathfrak{K}_{\pm,\mathfrak{R}}$. The quadruple $(V, \langle \cdot, \cdot \rangle, J_{\pm}, A)$ is said to be a *(para)-Kähler–Weyl curvature model* if $A \in \mathfrak{K}_{\pm,\mathfrak{W}}$. We refer to Brozos-Vázquez, Gilkey, and Nikčević [37] for the proof of Assertion (1) and to Brozos-Vázquez, Gilkey, and Merino [34] for the proof of Assertion (2) in the following result:

Theorem 1.14

1. *Every (para)-Kähler affine curvature model is geometrically realizable by a (para)-Kähler affine manifold.*

2. *Every (para)-Kähler Hermitian curvature model is geometrically realizable by a (para)-Kähler Hermitian manifold.*

The corresponding questions of geometric realizability for (para)-Kähler–Weyl curvature models will be examined in Chapter 4.

1.5   CURVATURE DECOMPOSITIONS

In this section, we will define the structure groups $\mathcal{O}$ (the orthogonal group), $\mathcal{U}_\pm$ (the (para)-unitary group), and $\mathcal{U}_\pm^\star$ (the $\star$-(para)-unitary group) and present the fundamental facts from representation theory that we shall need. We will also discuss results of Singer and Thorpe [194] giving the decomposition of the space of Riemannian algebraic curvature tensors $\mathfrak{R}$ as an $\mathcal{O}$ module, results of Higa [125, 126] giving the decomposition of the space of Weyl algebraic curvature tensors $\mathfrak{W}$ as an $\mathcal{O}$ module, and results of Tricerri and Vanhecke [200] giving the decomposition of $\mathfrak{R}$ as a $\mathcal{U}_\pm^\star$ module and the decomposition of the space of (para)-Kähler algebraic curvature tensors $\mathfrak{K}_{\pm,\mathfrak{R}}$ as a $\mathcal{U}_\pm^\star$ module. This will rise to the decomposition of $\mathfrak{W}$ as a $\mathcal{U}_\pm^\star$ module. As we shall not need the decomposition of $\mathfrak{K}_{\pm,\mathfrak{A}}$ as a $\mathcal{U}_\pm^\star$ module, we shall omit this decomposition and instead refer to the discussion in Brozos-Vázquez, Gilkey, and Nikčević [36] and also in Nikčević [159, 160]. We also refer to [22, 35, 38, 69, 79, 153, 198] for related work.

THE STRUCTURE GROUPS

If V is a vector space, let $\mathrm{GL} = \mathrm{GL}(V)$ denote the general linear group of linear transformations of V. If $(V, \langle \cdot, \cdot \rangle)$ is an inner product space, then the associated *orthogonal group* $\mathcal{O} = \mathcal{O}(V, \langle \cdot, \cdot \rangle)$ is given by:

$$\mathcal{O} := \{ T \in \mathrm{GL} : T^* \langle \cdot, \cdot \rangle = \langle \cdot, \cdot \rangle \}.$$

Let $(V, \langle \cdot, \cdot \rangle, J_\pm)$ be a (para)-Hermitian vector space. The associated structure groups are:

$$\mathcal{U}_\pm := \{ T \in \mathcal{O} : T J_\pm = J_\pm T \},$$
$$\mathcal{U}_\pm^\star := \{ T \in \mathcal{O} : T J_\pm = J_\pm T \text{ or } T J_\pm = -J_\pm T \}.$$

The group $\mathcal{U}_\pm$ is often called the *(para)-unitary group*. It is convenient to work with the $\mathbb{Z}_2$ extensions $\mathcal{U}_\pm^\star$ as we may then interchange the roles of $J_\pm$ and $-J_\pm$. Let $\chi(T) = \pm 1$ satisfy

$$J_\pm T = \chi(T) T J_\pm \quad \text{for} \quad T \in \mathcal{U}_\pm^\star. \tag{1.5.a}$$

Since $\chi(ST) = \chi(S)\chi(T)$, then we may regard χ as defining a $\mathbb{Z}_2$ valued character of $\mathcal{U}_\pm^\star$ or, alternatively, as being a 1-dimensional real representation of $\mathcal{U}_\pm^\star$. We can extend $\langle \cdot, \cdot \rangle$ to a natural non-degenerate inner product on $\otimes^k V$ and $\otimes^k V^*$. The following observation is fundamental in the subject:

Lemma 1.15   *Let $(V, \langle \cdot, \cdot \rangle, J_\pm)$ be a (para)-Hermitian vector space. Let $G \in \{\mathcal{O}, \mathcal{U}_-, \mathcal{U}_-^\star, \mathcal{U}_+^\star\}$ and let ξ be a G submodule of $\otimes^k V^*$. Then the restriction of the inner product on $\otimes^k V^*$ to ξ is non-degenerate.*

Proof. As this is a crucial observation, we shall give the proof. Let $\{e_i\}$ be an orthonormal basis for V and let $\{e^i\}$ be the associated dual basis for V^*. If $I = (i_1, \ldots, i_k)$ is a multi-index, set

$e^I = e^{i_1} \otimes \cdots \otimes e^{i_k}$. Then:

$$(e^I, e^J) := \langle e^{i_1}, e^{j_1} \rangle \cdots \langle e^{i_k}, e^{j_k} \rangle = \left\{ \begin{array}{ll} 0 & \text{if } I \neq J \\ \pm 1 & \text{if } I = J \end{array} \right\}. \tag{1.5.b}$$

Let $T e_i = \langle e_i, e_i \rangle \cdot e_i$ define an element $T \in \mathcal{O}$. Suppose ξ is an $\mathcal{O}$ invariant subspace of $\otimes^k V^*$. Decompose $\xi = \xi_+ \oplus \xi_-$ and decompose $\otimes^k V^* = Y_+ \oplus Y_-$ into the ± 1-eigenspaces of T. Since $T \in \mathcal{O}$, these decompositions are orthogonal direct sums. By Equation (1.5.b), Y_+ is spacelike and Y_- is timelike. Since $\xi_{\pm} \subset Y_{\pm}$, ξ_+ is spacelike and ξ_- is timelike; the Lemma now follows in this special case. If $G = \mathcal{U}_-$ or if $G = \mathcal{U}_-^{\star}$, then we can choose the orthonormal basis so that $J_- e_{2\nu-1} = e_{2\nu}$ and $J_- e_{2\nu} = -e_{2\nu-1}$; this additional normalization is crucial. Since $J_-^* \langle \cdot, \cdot \rangle = \langle \cdot, \cdot \rangle$, $J_- T = T J_-$. We then have $T \in G$ and the same argument pertains. Finally, suppose $G = \mathcal{U}_+^{\star}$. We can choose the basis so $J_+ e_{2\nu-1} = e_{2\nu}$, $J_+ e_{2\nu} = e_{2\nu-1}$, $e_{2\nu-1}$ is spacelike and $e_{2\nu}$ is timelike. We now have $T \in \mathcal{U}_+^{\star} - \mathcal{U}_+$. $\qquad \square$

We note that Lemma 1.15 fails for the group $G = \mathcal{U}_+$. For example, the ± 1-eigenspaces of J_+ are invariant under J_+ and are totally isotropic. We can combine Lemma 1.15 with same arguments as used in the positive definite setting to establish the following result; we omit details in the interests of brevity and instead refer to the discussion in Brozos-Vázquez, Gilkey, and Nikčević [37]:

Lemma 1.16   *Let $(V, \langle \cdot, \cdot \rangle, J_{\pm})$ be a (para)-Hermitian vector space. Let $G \in \{\mathcal{O}, \mathcal{U}_-, \mathcal{U}_-^{\star}, \mathcal{U}_+^{\star}\}$ and let ξ be a G submodule of $\otimes^k V^*$.*

1. *There exist G submodules η_i of ξ so that we may decompose $\xi = \eta_1 \oplus \cdots \oplus \eta_k$ as the orthogonal direct sum of irreducible G modules. The multiplicity with which an irreducible representation appears in ξ is independent of this decomposition.*

2. *If ξ_1 and ξ_2 are inequivalent irreducible submodules of ξ, then $\xi_1 \perp \xi_2$. If ξ_1 appears with multiplicity 1 in ξ and if η is any G submodule of ξ, then either $\xi_1 \subset \eta$ or else $\xi_1 \perp \eta$.*

We can illustrate Lemma 1.16 as follows. Let $(V, \langle \cdot, \cdot \rangle, J_{\pm})$ be a (para)-Hermitian vector space. We may decompose

$$\otimes^2 V^* = \Lambda^2 \oplus S^2$$

as the direct sum of the alternating and the symmetric bilinear forms. Let $\chi(T)$ defined by Equation (1.5.a) and let

$$\Omega_{\pm}(x, y) := \langle x, J_{\pm} y \rangle \tag{1.5.c}$$

be the *(para)-Kähler form*. Let $\mathbb{1}$ be the trivial module. If $T \in \mathcal{U}_{\pm}^{\star}$, then:

$$T^* \Omega_{\pm} = \chi(T) \Omega_{\pm}.$$

Let

$$\Lambda_\pm^2 = \Lambda_\pm^2(V^*) := \{\omega \in \Lambda^2 : J_\pm^* \omega = \pm\omega\}, \qquad \chi := \Omega_\pm \cdot \mathbb{R},$$
$$\Lambda_{\mp,0}^2 = \Lambda_{\pm,0}^2(V^*) := \{\omega \in \Lambda^2 : J_\pm^* \omega = \mp\omega, \ \omega \perp \Omega_\pm\},$$
$$S_\pm^2 = S_\pm^2(V^*) := \{\theta \in S^2 : J_\pm^* \theta = \pm\theta\}, \qquad \mathsf{I} := \langle \cdot, \cdot \rangle \cdot \mathbb{R}, \qquad \text{(1.5.d)}$$
$$S_{\mp,0}^2 = S_{\pm,0}^2(V^*) := \{\theta \in S^2 : J_\pm^* \theta = \mp\theta, \ \theta \perp \langle \cdot, \cdot \rangle\}.$$

Lemma 1.17   *Let $(V, \langle \cdot, \cdot \rangle, J_\pm)$ be a (para)-Hermitian vector space. We have the following decomposition of Λ^2, S^2, and $\otimes^2 V^*$ into inequivalent and irreducible $\mathcal{U}_\pm^\star$ modules:*

$$\Lambda^2 = \Lambda_\pm^2 \oplus \chi \oplus \Lambda_{\mp,0}^2, \qquad S^2 = S_\pm^2 \oplus \mathsf{I} \oplus S_{\mp,0}^2,$$
$$\otimes^2 V^* = \Lambda_\pm^2 \oplus \chi \oplus \Lambda_{\mp,0}^2 \oplus S_\pm^2 \oplus \mathsf{I} \oplus S_{\mp,0}^2.$$

We note that $\Lambda_{\mp,0}^2$ and $S_{\mp,0}^2$ are isomorphic $\mathcal{U}_\pm$ modules, that $\Lambda_{\mp,0}^2$ is isomorphic to $S_{\mp,0}^2 \otimes \chi$ as a $\mathcal{U}_\pm^\star$ module, that χ is isomorphic to I as a $\mathcal{U}_\pm$ module, and that Λ_+^2 is not an irreducible $\mathcal{U}_+$ module.

ALMOST (PARA)-HERMITIAN STRUCTURES

We now present results of Brozos-Vázquez *et al.* [30] giving the classification of (para)-Hermitian structures. Let $(M, g, J_\pm)$ be an almost (para)-Hermitian manifold. Let ∇ be the Levi-Civita connection of g. The associated (para)-Kähler form $\Omega_\pm$ and the covariant derivative $\nabla\Omega_\pm$ are given by:

$$\Omega_\pm(X, Y) := g(X, J_\pm Y),$$
$$\nabla\Omega_\pm(X, Y; Z) = Zg(X, J_\pm Y) - g(\nabla_Z X, J_\pm Y) - g(X, J_\pm \nabla_Z Y).$$

One has the symmetries:

$$\nabla\Omega_\pm(X, Y; Z) = -\nabla\Omega_\pm(Y, X; Z) \text{ and } \nabla\Omega_\pm(X, Y; Z) = \pm\nabla\Omega_\pm(J_\pm X, J_\pm Y; Z). \quad \text{(1.5.e)}$$

It is convenient to work in an algebraic setting similar to that which we will use subsequently in discussing curvature models. Let $(V, \langle \cdot, \cdot \rangle, J_\pm)$ be a (para)-Hermitian vector space. Motivated by Equation (1.5.e), we define:

$$\mathcal{H}_\pm := \{H_\pm \in \otimes^3 V^* : H_\pm(x, y; z) = -H_\pm(y, x; z) \quad \text{and}$$
$$H_\pm(J_\pm x, J_\pm y; z) = \pm H_\pm(x, y; z), \ \forall\, x, y, z\},$$
$$\mathcal{U}_{3,\pm} := \{H_\pm \in \mathcal{H}_\pm : H_\pm(x, y; z) = \mp H_\pm(x, J_\pm y; J_\pm z), \ \forall\, x, y, z\}.$$

Let $\varepsilon_{ij} := \langle e_i, e_j \rangle$ give the components of the inner product relative to some basis $\{e_i\}$ for V. We define:

$(\tau_1 H)(x) := \varepsilon^{ij} H(x, e_i; e_j)$ for $H \in \otimes^3 V^*$,

$\nu_\pm(\phi)(x, y; z) := \phi(J_\pm x)\langle y, z \rangle - \phi(J_\pm y)\langle x, z \rangle$

$\qquad\qquad + \phi(x)\langle J_\pm y, z \rangle - \phi(y)\langle J_\pm x, z \rangle$ for $\phi \in V^*$,

$W_{1,\pm} := \{H_\pm \in \mathcal{H}_\pm : H_\pm(x, y; z) + H_\pm(x, z; y) = 0, \; \forall x, y, z\}$,

$W_{2,\pm} := \{H_\pm \in \mathcal{H}_\pm : H_\pm(x, y; z) + H_\pm(y, z; x) + H_\pm(z, x; y) = 0, \; \forall x, y, z\}$,

$W_{3,\pm} := \mathcal{U}_{3,\pm} \cap \ker\{\tau_1\}$,

$W_{4,\pm} := \mathrm{Range}\{\nu_\pm\}$.

In the geometric setting, if $(M, g, J_\pm)$ is a (para)-Hermitian manifold (i.e., $J_\pm$ is integrable), then $\nabla\Omega_\pm \in \mathcal{U}_{3,\pm}$. The following result in Brozos-Vázquez *et al.* [30] is based on a decomposition of $\mathcal{H}_\pm$, which extends the Gray–Hervella classification [118] of almost Hermitian structures in the positive definite context to the indefinite setting and to the para-complex setting:

Theorem 1.18   *Let $m \geq 6$. We have a direct sum orthogonal decomposition of $\mathcal{H}_\pm$ and of $\mathcal{U}_{3,\pm}$ into irreducible inequivalent $\mathcal{U}_\pm^\star$ modules in the form:*

$$\mathcal{H}_\pm = W_{1,\pm} \oplus W_{2,\pm} \oplus W_{3,\pm} \oplus W_{4,\pm} \quad \text{and} \quad \mathcal{U}_{3,\pm} = W_{3,\pm} \oplus W_{4,\pm} \, .$$

One obtains the corresponding decompositions if $m = 4$, by setting $W_{1,\pm} = 0$ and $W_{3,\pm} = 0$.

The fact that there are only two components in dimension four has important implications in (para)-Kähler–Weyl geometry and is one of the reasons why the 4-dimensional setting is exceptional in that geometry.

It was shown in Brozos-Vázquez *et al.* [30] that every element of $\mathcal{H}_\pm$ and of $\mathcal{U}_{3,\pm}$ is geometrically realizable in an appropriate context. One can, however, focus instead on the precise nature of the classes involved. Restricting to the complex setting, we say that (M, g, J_-) is a ξ-manifold if $\nabla\Omega_-$ belongs to ξ at every point of the manifold, where ξ is a $\mathcal{U}_-^\star$ submodule of $\mathcal{H}_-$, which is minimal with this property. This gives rise to the celebrated 16 classes of almost Hermitian manifolds in the positive definite setting. Many of these classes have geometrical meanings which have been extensively investigated. For example:

1. $\xi = \{0\}$ defines the class of Kähler manifolds.

2. $\xi = W_{1,-}$ defines the class of nearly Kähler manifolds.

3. $\xi = W_{2,-}$ defines the class of almost Kähler manifolds.

4. $\xi = W_{3,-}$ defines the class of Hermitian semi-Kähler manifolds.

5. $\xi = W_{1,-} \oplus W_{2,-}$ defines the class of quasi-Kähler manifolds.

6. $\xi = W_{3,-} \oplus W_{4,-} = \mathcal{U}_{3,-}$ defines the class of Hermitian manifolds.

7. $\xi = W_{1,-} \oplus W_{2,-} \oplus W_{3,-}$ defines the class of semi-Kähler manifolds.

8. $\xi = \mathcal{H}_-$ defines the class of almost Hermitian manifolds.

SCALAR INVARIANTS

Let $(V, \langle \cdot, \cdot \rangle, J_\pm)$ be a (para)-Hermitian vector space and let ξ be a G submodule of $\otimes^k V^*$ where $G \in \{\mathcal{O}, \mathcal{U}_-, \mathcal{U}_-^\star, \mathcal{U}_+^\star\}$. We say that a linear map Ξ from ξ to $\mathbb{R}$ is a *scalar invariant* if $\Xi(g \cdot v) = \Xi(v)$ for every $v \in \xi$ and for every $g \in G$; let $\mathcal{I}^G(\xi)$ be the vector space of all such invariants. The inner product $\langle \cdot, \cdot \rangle$ extends naturally to an inner product $\langle \cdot, \cdot \rangle$ on ξ, which is non-degenerate by Lemma 1.15. Let $\mathrm{Hom}_G\{\xi\}$ be the set of *intertwining operators*, i.e., linear maps T of ξ to ξ so that $g^{-1}Tg = T$ for all $g \in G$. Let θ be a bilinear form on ξ, i.e., a linear map from $\xi \otimes \xi$ to $\mathbb{R}$. Define a linear map T_θ from ξ to ξ, which is characterized by the property $\theta(v_1, v_2) := \langle v_1, T_\theta v_2 \rangle$.

Lemma 1.19   *The map $\theta \to T_\theta$ identifies $\mathcal{I}^G(\xi \otimes \xi)$ with $\mathrm{Hom}_G\{\xi\}$. Consequently, if $\dim(\mathcal{I}^G(\xi \otimes \xi)) = 1$, then any intertwining operator $T \in \mathrm{Hom}_G\{\xi\}$ is a scalar multiple of the identity.*

Proof. A linear invariant on $\xi \otimes \xi$ can be regarded as a bilinear form on ξ or, equivalently, as an element of $\otimes^2 \xi^*$. We have

$$\theta(gv_1, gv_2) = \langle gv_1, T_\theta gv_2 \rangle = \langle v_1, g^* T_\theta gv_2 \rangle = \langle v_1, g^{-1} T_\theta gv_2 \rangle.$$

Thus, $\theta \in \mathcal{I}(\xi \otimes \xi)$ is a linear invariant if and only if $g^{-1} T_\theta g = T_\theta$ or, equivalently, if T_θ belongs to $\mathrm{Hom}_G\{\xi\}$. $\qquad \square$

Weyl [210, pages 53 and 66] gives a spanning set if $G = \mathcal{O}$ is the orthogonal group; the corresponding result for the unitary group $\mathcal{U}_-$ in the positive definite Hermitian setting follows from work of Fukami [84] and of Iwahori [136]; the extension to the groups $\mathcal{U}_\pm^\star$ in general is straightforward—see, for example, Brozos-Vázquez, Gilkey, and Nikčević [37].

We discuss this spanning set. All invariants arise by using either the inner product or the (para)-Kähler form to contract indices in pairs; invariants of $\mathcal{U}_\pm^\star$ arise when the (para)-Kähler form appears an even number of times. It is worth being a bit more formal about this. Let $(V, \langle \cdot, \cdot \rangle, J_\pm)$ be a (para)-Hermitian vector space. Let $h = \langle \cdot, \cdot \rangle$ and let $\Omega_{\pm,ij}$ be the components of the (para)-Kähler form. If $\{e_i\}$ is any basis for V, let $h_{ij} := \langle e_i, e_j \rangle$ give the components of h relative to this basis. The inverse matrix $h^{ij} = \langle e^i, e^j \rangle$ gives the components of the dual inner product on V^*. Let $\Theta \in \otimes^{2k} V^*$. We may expand

$$\Theta = \Theta_{i_1 \dots i_{2k}} e^{i_1} \otimes \cdots \otimes e^{i_{2k}}.$$

Let π be a permutation of the set $\{1, \dots, 2k\}$. Let $\kappa_0 := h$, let $\kappa_1 := \Omega_\pm$, and let $\vec{a}$ be a sequence of 0's and 1's. Define:

$$\psi_{\pi, \vec{a}}(\Theta) := \kappa_{a_1}^{i_{\pi(1)} i_{\pi(2)}} \dots \kappa_{a_k}^{i_{\pi(2k-1)} i_{\pi(2k)}} \Theta_{i_1 \dots i_{2k}}.$$

Let $n(\vec{a})$ be the number of times $a_i = 1$. One then has:

Lemma 1.20 *If $(V, \langle \cdot, \cdot \rangle, J_\pm)$ is a (para)-Hermitian vector space and if ξ is a $\mathcal{U}_\pm^\star$ submodule of $\otimes^{2k} V^*$, then $\mathcal{I}^{\mathcal{U}_\pm^\star}(\xi) = \operatorname{Span}_{n(\vec{a}) \ even}\{\psi_{\pi, \vec{a}}\}$.*

Applying this formalism in the geometric setting to the curvature tensor and to the covariant derivatives of the curvature tensor yields scalar invariants of the metric. Let $\mathcal{I}_n$ be the space of invariants which are homogeneous of order n in the jets of the metric. Let ";" denotes covariant differentiation. One has, see for example Gilkey [94], that:

Lemma 1.21 *Let (M, g) be a Riemannian manifold. Let $\{e_i\}$ be a local frame orthonormal field for the manifold. Then*

$$\mathcal{I}_0 = \operatorname{Span}\{1\},$$
$$\mathcal{I}_2 = \operatorname{Span}\{R_{ijji}\},$$
$$\mathcal{I}_4 = \operatorname{Span}\{R_{ijji;kk}, R_{ijji}R_{k\ell\ell k}, R_{ijjk}R_{i\ell\ell k}, R_{ijk\ell}R_{ijk\ell}\},$$
$$\mathcal{I}_6 = \operatorname{Span}\{R_{ijji;kk\ell\ell}, R_{ijji;k}R_{\ell nn\ell;k}, R_{aija;k}R_{bijb;k}, R_{ajka;n}R_{bjnb;k},$$
$$R_{ijk\ell;n}R_{ijk\ell;n}, R_{ijji}R_{k\ell\ell k;nn}, R_{ajka}R_{bjkb;nn}, R_{ajka}R_{bjnb;kn},$$
$$R_{ijk\ell}R_{ijk\ell;nn}, R_{ijji}R_{k\ell\ell k}R_{abba}, R_{ijji}R_{ajka}R_{bjkb}, R_{ijji}R_{abcd}R_{abcd},$$
$$R_{ajka}R_{bjnb}R_{cknc}, R_{aija}R_{bk\ell b}R_{ikj\ell}, R_{ajka}R_{jn\ell i}R_{kn\ell i}, R_{ijkn}R_{ij\ell p}R_{kn\ell p},$$
$$R_{ijkn}R_{i\ell kp}R_{j\ell np}\}$$

These invariants are linearly independent if $m \geq 6$, but in lower dimensions there are relations amongst these invariants called *universal curvature identities* that arise from Weyl's *second theorem of invariants*; we refer to the discussion in Gilkey, Park, and Sekigawa [111, 112] for further details. There is a similar formalism in the pseudo-Riemannian setting where one has to be a bit more careful since $g(e_i, e_i) = \pm 1$.

A Riemannian manifold is locally homogeneous if and only if all the scalar invariants up to order $\frac{1}{2}m(m-1)$ are constant and the local geometry is determined by these invariants (see Prüfer, Tricerri, and Vanhecke [183]). This fails in the pseudo-Riemannian setting as there are manifolds all of whose scalar invariants vanish—see, for example, the discussion in Alcolado *et al.* [4], in Coley *et al.* [58], and in Pravda *et al.* [182]. Such manifolds are called *vanishing scalar invariants* manifolds or simply *VSI* manifolds. We complete our discussion of elementary representation theory with the following result (see, for example, the discussion in Brozos-Vázquez, Gilkey, and Nikčević [37]):

Lemma 1.22 *Let $(V, \langle \cdot, \cdot \rangle, J_\pm)$ be a (para)-Hermitian vector space.*

1. $\dim(\operatorname{Hom}_{\mathcal{U}_-^\star}\{\Lambda_-^2\}) = 1$ *and* $\dim(\operatorname{Hom}_{\mathcal{U}_+^\star}\{\Lambda_+^2\}) = 1$.

2. *Let $\xi(a, b) := \{(a\theta, b\theta)\}_{\theta \in \Lambda_\pm^2} \subset \Lambda_\pm^2 \oplus \Lambda_\pm^2$ for $(a, b) \neq (0, 0)$. If ξ is a non-trivial proper $\mathcal{U}_\pm^\star$ submodule of $\Lambda_\pm^2 \oplus \Lambda_\pm^2$, then there exists $0 \neq (a, b) \in \mathbb{R}^2$ so $\xi = \xi(a, b)$.*

Proof. We prove Assertion (1) as follows. We have, by definition, that:

$$\Lambda_\pm^2 \otimes \Lambda_\pm^2 = \{\theta \in \otimes^4 V^* : \theta(x, y, z, w) = -\theta(y, x, z, w) = -\theta(x, y, w, z)$$
$$\text{and } \theta(x, y, z, w) = \pm\theta(J_\pm x, J_\pm y, z, w) = \pm\theta(x, y, J_\pm z, J_\pm w) \, \forall x, y, z, w\}.$$

Let $\{e_i\}$ be a basis for V and let $\varepsilon_{ij} = \langle e_i, e_j \rangle$. By Lemma 1.20, any $\mathcal{U}_\pm^\star$ linear invariant of $\Lambda_\pm^2 \otimes \Lambda_\pm^2$ is a linear combination of the invariants:

$$\Psi_1(\theta) = \varepsilon^{ij}\varepsilon^{k\ell}\theta_{ijk\ell}, \quad \Psi_2(\theta) = \varepsilon^{ik}\varepsilon^{j\ell}\theta_{ijk\ell}, \quad \Psi_3(\theta) = \varepsilon^{i\ell}\varepsilon^{jk}\theta_{ijk\ell},$$
$$\Psi_4(\theta) = \Omega_\pm^{ij}\Omega_\pm^{k\ell}\theta_{ijk\ell}, \quad \Psi_5(\theta) = \Omega_\pm^{ik}\Omega_\pm^{j\ell}\theta_{ijk\ell}, \quad \Psi_6(\theta) = \Omega_\pm^{i\ell}\Omega_\pm^{jk}\theta_{ijk\ell}.$$

Since θ is skew-symmetric in the first two indices and since ε is symmetric,

$$\Psi_1(\theta) = 0, \quad \Psi_2(\theta) = -\Psi_3(\theta), \quad \Psi_5(\theta) = -\Psi_6(\theta).$$

We examine cases. We can decouple the bases. We consider 4 different bases $\{a_i\}, \{b_i\}, \{c_i\}$, and $\{d_i\}$ and the corresponding dual bases $\{a^i\}, \{b^i\}, \{c^i\}$, and $\{d^i\}$. We may then express

$$\Psi_4(\theta) = \sum_{i,j,k,\ell} \langle a^i, J_\pm b^j \rangle \langle c^k, J_\pm d^\ell \rangle \theta(a_i, b_j, c_k, d_\ell).$$

We set $a_i = J_\pm e_i, b_i = e_i, c_i = J_\pm e_i$, and $d_i = e_i$. This is equivalent to interchanging the order of summation and occasionally changing the sign. We show $\Psi_4(\theta) = 0$ by computing:

$$\Psi_4(\theta) = \langle J_\pm e^i, J_\pm e^j \rangle \langle J_\pm e^k, J_\pm e^\ell \rangle \theta(J_\pm e_i, e_j, J_\pm e_k, e_\ell)$$
$$= \langle e^i, e^j \rangle \langle e^k, e^\ell \rangle \theta(J_\pm e_i, e_j, J_\pm e_k, e_\ell)$$
$$= \pm\langle e^i, e^j \rangle \langle e^k, e^\ell \rangle \theta(J_\pm J_\pm e_i, J_\pm e_j, J_\pm e_k, e_\ell)$$
$$= \langle e^i, e^j \rangle \langle e^k, e^\ell \rangle \theta(e_i, J_\pm e_j, J_\pm e_k, e_\ell)$$
$$= -\langle e^i, e^j \rangle \langle e^k, e^\ell \rangle \theta(J_\pm e_j, e_i, J_\pm e_k, e_\ell)$$
$$= -\Psi_4(\theta).$$

Similarly, setting $a_i = J_\pm e_i, b_i = J_\pm e_i$, and $c_i = d_i = e_i$, we express:

$$\Psi_5(\theta) = \langle J_\pm e^i, J_\pm e^k \rangle \langle J_\pm e^j, J_\pm e^\ell \rangle \theta(J_\pm e^i, J_\pm e^j, e^k, e^\ell)$$
$$= \pm\langle e^i, e^k \rangle \langle e^j, e^\ell \rangle \theta(e_i, e_j, e_k, e_\ell) = \pm\Psi_2(\theta).$$

This shows $\dim(\mathcal{I}^{\mathcal{U}_\pm^\star}(\Lambda_\pm^2 \otimes \Lambda_\pm^2)) = 1$; Assertion (1) now follows from Lemma 1.19.

Let ξ be a proper $\mathcal{U}_\pm^\star$ submodule of $\Lambda_\pm^2 \oplus \Lambda_\pm^2$. Let π_1 (resp. π_2) be the projection on the first (resp. on the second) factor. Since ξ is non-trivial, we may assume, without loss of generality, that $\pi_1(\xi) \neq \{0\}$; since ξ is a proper submodule, ξ is necessarily irreducible, and hence π_1 is an isomorphism. If $\pi_2(\xi) = 0$, then $\xi = \xi(1, 0)$. Thus, we may assume that $\pi_2 \neq 0$ and consequently $T := \pi_2^{-1}\pi_1$ is a non-trivial $\mathcal{U}_\pm^\star$ equivariant map of $\Lambda_\pm^2$. By Assertion (1), any intertwining operator is a scalar multiple of the identity. Consequently, $T = b\,\mathrm{Id}$ and $\xi = \xi(1, b)$. $\qquad\square$

CONFORMAL EQUIVALENCE

Two pseudo-Riemannian manifolds (M_1, g_1) and (M_2, g_2) are said to be *(locally) conformally equivalent* if there is a (local) diffeomorphism $\varphi : M_1 \to M_2$ such that $\varphi^* g_2 = e^{2f} g_1$ for some smooth *conformal factor* f, which is defined (locally) on M. The *Weyl conformal curvature tensor* W plays a central role in these geometries. It forms one of the three components of the curvature tensor decomposition of Singer and Thorpe [194] that we shall discuss presently in Theorem 1.24. It is defined by setting:

$$W(X, Y, Z, T) := R(X, Y, Z, T) + \tfrac{1}{(m-1)(m-2)}\tau\{g(Y, Z)g(X, T) - g(X, Z)g(Y, T)\}$$
$$- \tfrac{1}{m-2}\{g(Y, Z)\rho(X, T) - g(X, Z)\rho(Y, T)\} \quad (1.5.f)$$
$$- \tfrac{1}{m-2}\{g(X, T)\rho(Y, Z) - g(Y, T)\rho(X, Z)\},$$

where ρ is the Ricci tensor and $\tau = \mathrm{Tr}\{\rho\}$ is the scalar curvature. Note that the Weyl conformal curvature tensor W is not defined for $m = 2$ and that W vanishes identically if $m = 3$.

The Weyl conformal curvature tensor defined above is invariant by conformal transformations, i.e., $^{g_1}W = e^{2f}\, {}^{g_2}W$ (or, equivalently, the associated Weyl curvature operators satisfy $^{g_1}\mathcal{W} = {}^{g_2}\mathcal{W}$) for any two conformally equivalent metrics. A converse implication holds true under certain conditions that depend on the signature of the metric, the dimension, and the non-degeneracy of the Weyl curvature operator—we refer to Hall [120] for further details. A pseudo-Riemannian manifold (M, g) is said to be *locally conformally flat* if it is locally conformally equivalent to a flat manifold. If (M, g) is locally conformally flat, then

$$W = 0 \text{ and } (\nabla_X\rho)(Y, Z) = (\nabla_Y\rho)(X, Z) \text{ for all } X, Y, Z \in C^\infty(TM).$$

The following converse is due to Weyl [207, 208] and Schouten [191]—see also Hertrich-Jeromin [124].

Lemma 1.23 *Let (M, g) be a pseudo-Riemannian manifold. Then (M, g) is locally conformally flat if and only if one of the following conditions holds:*

1. *$m \geq 4$ and $^g W = 0$.*

2. *$m = 3$ and $(\nabla_X\rho)(Y, Z) = (\nabla_Y\rho)(X, Z)$ for all vector fields X, Y, Z on M.*

Finally, observe that the Ricci tensor of any locally conformally flat pseudo-Riemannian manifold (M, g) satisfies the identity in Assertion (2) above.

THE SINGER–THORPE AND HIGA DECOMPOSITIONS

We now examine the $\mathcal{O}$ module structure of $\mathfrak{R}$ and $\mathfrak{W}$. Let $(V, \langle \cdot, \cdot \rangle)$ be an inner product space. Let $S^2 = S^2(V^*)$, and let

$$S_0^2 := \{\theta \in S^2 : \theta \perp \langle \cdot, \cdot \rangle\} \quad \text{and} \quad \mathfrak{C} := \ker\{\rho\} \cap \mathfrak{R}$$

be the $\mathcal{O}$ modules of trace-free symmetric 2-cotensors and *Weyl conformal curvature tensors*, respectively. We refer to Singer and Thorpe [194] for the proof of Assertion (1) and to Higa [125, 126] for the proof of Assertion (2) in the following result:

Theorem 1.24   *Let $m \geq 4$.*

1. *We may decompose $\mathfrak{R} = \mathbf{1} \oplus S_0^2 \oplus \mathfrak{C}$ as the orthogonal direct sum of irreducible and inequivalent $\mathcal{O}$ modules.*

2. *We may decompose $\mathfrak{W} = \mathbf{1} \oplus S_0^2 \oplus \mathfrak{C} \oplus \mathfrak{P}$ as the orthogonal direct sum of irreducible and inequivalent $\mathcal{O}$ modules. Here, ρ_a provides an $\mathcal{O}$ module isomorphism from $\mathfrak{P}$ to Λ^2 with the inverse embedding $\Xi : \Lambda^2 \overset{\approx}{\longrightarrow} \mathfrak{P} \subset \mathfrak{W}$ being given by:*

$$\Xi(\psi)(x, y, z, w) := 2\psi(x, y)\langle z, w\rangle + \psi(x, z)\langle y, w\rangle - \psi(y, z)\langle x, w\rangle$$
$$- \psi(x, w)\langle y, z\rangle + \psi(y, w)\langle x, z\rangle .$$

The space $\mathfrak{C}$ consists of the Weyl conformal curvature tensors defined in Equation (1.5.f). The trivial module $\mathbf{1}$ is detected by the scalar curvature, and the module S_0^2 is detected by the trace-free Ricci tensor.

THE TRICERRI–VANHECKE DECOMPOSITIONS

The following decompositions of $\mathfrak{R}$ and $\mathfrak{K}_{\pm,\mathfrak{R}}$ as $\mathcal{U}_-$ modules were given in Mori [156], in Sitaramayya [195], and in Tricerri and Vanhecke [200] in the positive definite setting; they extend easily to the more general context (see Brozos-Vázquez, Gilkey, and Merino [34] and Brozos-Vázquez, Gilkey, and Nikčević [37]). The decomposition of $\mathfrak{W}$ as a $\mathcal{U}_{\pm}^{\star}$ module then follows from Lemma 1.17 and Theorem 1.24.

Theorem 1.25   *Let $(V, \langle\cdot,\cdot\rangle, J_{\pm})$ be a (para)-Hermitian vector space. We have the following decompositions of $\mathfrak{R}(V)$, $\mathfrak{K}_{\pm,\mathfrak{R}}(V)$, and $\mathfrak{W}(V)$ as $\mathcal{U}_{\pm}^{\star}$ modules:*

$$\mathfrak{R}(V) = W_{\pm,1} \oplus \cdots \oplus W_{\pm,10}, \qquad \mathfrak{K}_{\pm,\mathfrak{R}}(V) = W_{\pm,1} \oplus W_{\pm,2} \oplus W_{\pm,3},$$
$$\mathfrak{W}(V) = W_{\pm,1} \oplus \cdots \oplus W_{\pm,13}.$$

If $m = 4$, we omit the modules $\{W_{\pm,5}, W_{\pm,6}, W_{\pm,10}\}$. If $m = 6$, we omit the module $W_{\pm,6}$. The decompositions given above are then into irreducible $\mathcal{U}_{\pm}^{\star}$ modules. We have $\mathcal{U}_{\pm}^{\star}$ module isomorphisms: $W_{\pm,1} \approx W_{\pm,4} \approx \mathbf{1}$, $W_{\pm,2} \approx W_{\pm,5} \approx S_{\mp,0}^2$, $W_{\pm,9} \approx W_{\pm,13} \approx \Lambda_{\pm}^2$. $W_{\pm,8} \approx S_{\pm}^2$, $W_{\pm,11} \approx \chi$, and $W_{\pm,12} \approx \Lambda_{\mp,0}^2$. With the exception of these isomorphisms, the modules $W_{\pm,i}$ are inequivalent $\mathcal{U}_{\pm}^{\star}$ modules. The isomorphism Ψ from $\Lambda_{\pm}^2$ to $W_{\pm,9}$ is given by setting:

$$\Psi(\psi)(x, y, z, w) := 2\langle x, J_\pm y\rangle \psi(z, J_\pm w) + 2\langle z, J_\pm w\rangle \psi(x, J_\pm y)$$
$$+\langle x, J_\pm z\rangle \psi(y, J_\pm w) + \langle y, J_\pm w\rangle \psi(x, J_\pm z)$$
$$-\langle x, J_\pm w\rangle \psi(y, J_\pm z) - \langle y, J_\pm z\rangle \psi(x, J_\pm w).$$

Remark 1.26   The *Bochner curvature tensor* of an almost (para)-Hermitian manifold $(M, g, J_\pm)$ may be defined projecting $\mathfrak{R}(V)$ to the subspace $W_{\pm,3} \oplus W_{\pm,6} \oplus W_{\pm,7} \oplus W_{\pm,10}$ (see Bryant [39] or Tricerri and Vanhecke [200]). This tensor is analogous to the Weyl conformal curvature tensor defined in Equation (1.5.f).

1.6   WALKER STRUCTURES

In this section, we will review the main results concerning Walker structures that we shall need subsequently. Holonomy, parallel transport, and isometric decompositions are treated, as are parallel plane fields and Walker coordinates. Subsequently, in Section 1.7, we treat a special class of Walker metrics called Riemannian extensions. We also discuss deformed Riemannian extensions and modified Riemannian extensions. The field is a vast one and we can only give a few references on the subject [3, 43, 45, 52, 68, 75, 89, 152, 170, 177, 203, 211]; see also Law and Matsushita [148] for a spinor approach.

THE HOLONOMY GROUP AND ISOMETRIC DECOMPOSITIONS

Let $\alpha : [a, b] \to M$ be a smooth curve in a connected affine manifold (M, ∇). For each tangent vector v in the tangent space to M at the initial point of the curve $\alpha(a) = P$, let $v(t)$ be defined by parallel transport along α; $v(t)$ is the vector field along α given by solving the equation

$$\nabla_{\dot\alpha(t)} v(t) = 0 \text{ with initial condition } v(a) = v.$$

If α is a closed curve, then the map $v(a) \to v(b)$ given by the parallel transport around α defines a linear isomorphism

$$L_\alpha : T_P M \to T_P M ,$$

which is called *holonomy*. The set of all such linear maps forms a group called the *holonomy group of the connection*. Since M is assumed to be connected, the holonomy groups corresponding to different points of M are all isomorphic; if β is any curve from P to Q, then L_β provides an isomorphism between the holonomy group at P and the holonomy group at Q. Consequently, the role of the basepoint P is usually suppressed. There is a vast literature on the subject—we cite a few representative examples [5, 11, 76, 80, 121, 150].

The holonomy group is a closed subgroup of the general linear group $GL(T_P M)$ of the tangent bundle at P. Consequently, the holonomy group is a Lie group. When one uses the Levi-Civita connection of a pseudo-Riemannian metric, the holonomy group is a subgroup of the orthogonal group $\mathcal{O}(T_P M, g_P)$, since parallel transport is realized by isometries, and we refer to it as the *holonomy group of the pseudo-Riemannian manifold*.

In the Riemannian setting the holonomy group acts completely *reducibly*. This means that there is an orthogonal direct sum decomposition

$$T_P M = V_1 \oplus \cdots \oplus V_\ell \,,$$

where each V_i is invariant under the holonomy group and where V_i contains no non-trivial invariant subspaces. In the higher signature setting, however, the situation is more complicated. If V_i is an invariant subspace, there need not exist a complementary invariant subspace. One says that the holonomy group acts *indecomposably* and that the manifold is *indecomposable* if the metric is degenerate on any invariant proper subspace. In the Riemannian setting, indecomposability is equivalent to irreducibility.

The holonomy group of the product of pseudo-Riemannian manifolds is the product of the holonomy groups of these manifolds. Furthermore, a converse of this statement is true in the following sense. Suppose that (M, g) is a connected pseudo-Riemannian manifold such that the tangent space at a single point (and hence at every point) admits an orthogonal direct sum decomposition into non-degenerate subspaces which are invariant under the holonomy representation. In this setting, (M, g) is locally isometric to a product of pseudo-Riemannian manifolds corresponding to the invariant subspaces. Moreover, the holonomy group is the product of the groups acting on the corresponding invariant subspaces. A global version of this statement, under the assumption that the manifold is simply connected and complete, was established by de Rham [184] for Riemannian manifolds and by Wu [214] in the arbitrary signature case:

Theorem 1.27   *Any simply connected complete pseudo-Riemannian manifold (M, g) is isometric to a product of simply connected complete pseudo-Riemannian manifolds, one of which can be flat and the others have an indecomposably acting holonomy group. Moreover, the holonomy group of (M, g) is the product of these indecomposably acting holonomy groups.*

PARALLEL PLANE FIELDS

Let (M, g) be a pseudo-Riemannian manifold and let $\mathfrak{D}$ be a distribution, i.e., a smooth subbundle of the tangent space. We say that $\mathfrak{D}$ is a *parallel plane field* if

$$\nabla_X Y \in \mathfrak{D} \text{ for all } X \in C^\infty(TM) \text{ and for all } Y \in C^\infty(\mathfrak{D}) \,.$$

The distribution $\mathfrak{D}$ is said to be a *null plane field* if $\mathfrak{D}$ is *totally isotropic*, i.e., if

$$g(X, Y) = 0 \text{ for all } X \in C^\infty(\mathfrak{D}) \text{ and } Y \in C^\infty(\mathfrak{D}) \,.$$

The existence of a null parallel plane field $\mathcal{D}$ in a pseudo-Riemannian manifold influences the curvature. We refer to the discussion in Derdzinski and Roter [68] for the proof of the following result:

Lemma 1.28   *Let* $\mathcal{M} := (M, g)$ *be a pseudo-Riemannian manifold admitting a null parallel plane field* $\mathcal{D}$. *Then* $R(\mathcal{D}, \mathcal{D}^{\perp}, \cdot, \cdot) = 0$, $R(\mathcal{D}, \mathcal{D}, \cdot, \cdot) = 0$, *and* $R(\mathcal{D}^{\perp}, \mathcal{D}^{\perp}, \mathcal{D}, \cdot) = 0$.

WALKER METRICS AND WALKER COORDINATES

For indefinite metrics, there exists the possibility that some of the factors in Theorem 1.27 are indecomposable, but not irreducible. This means that one of the factors could contain a proper null invariant subspace. Indecomposable Lorentzian metrics were initially investigated by Brinkmann [28], while the general pseudo-Riemannian case was first considered by Walker [204]. In all cases, indecomposability is equivalently described by the existence of a null parallel plane field. Following the notation in Brozos-Vázquez *et al.* [31], and motivated by this seminal work of Walker, a pseudo-Riemannian manifold (M, g) that admits a non-trivial null parallel plane field $\mathcal{D}$ is said to be a *Walker manifold*. Such metrics only arise in the pseudo-Riemannian setting. They form the underlying structure of many pseudo-Riemannian situations with no Riemannian counterpart. We refer to Brozos-Vázquez *et al.* [31] for more information on this family of metrics.

By imposing some additional curvature identities of the form given in Lemma 1.28, one can specialize the *Walker structure* under consideration. This is the case, for example, for the *pp-waves* which are locally characterized by the identity $R(\mathcal{D}^{\perp}, \mathcal{D}^{\perp}, \cdot, \cdot) = 0$ and by the fact that the range of the Ricci operator is totally isotropic (see Leistner [150]). Note that $T_P M \neq \mathcal{D} \oplus \mathcal{D}^{\perp}$ since $\mathcal{D}$ is a null distribution. There is a vast literature concerning pp-waves and we cite a few representative examples here [1, 139, 149, 155].

The existence of adapted coordinates $(u, v, x^1, \ldots, x^{m-2})$ on a Lorentzian manifold (M, g), admitting a null parallel line field, was established by Brinkmann [28]. In those coordinates the metric g has the form:

$$g = 2\, du \circ dv + f\, du \circ du + a_i\, du \circ dx^i + g_{ij}\, dx^i \circ dx^j \text{ where } \partial_v g_{ij} = \partial_v a_i = 0 .$$

Moreover $\partial_v f = 0$ if and only if the null parallel line field $\mathcal{D} = \text{Span}\{\partial_v\}$ is spanned by a parallel vector field, in which case the coordinates can be chosen so that $a_i = 0$ and so that $f = 0$. Walker [204] generalized this result to arbitrary pseudo-Riemannian manifolds (M, g) that admit a null parallel plane field $\mathcal{D}$ of dimension r as follows:

Theorem 1.29   *If* (M, g) *is an m-dimensional pseudo-Riemannian manifold which admits a null parallel plane field* $\mathcal{D}$ *of dimension* r, *then there exist local coordinates* $(x^1, \ldots, x^{m-r}, x^{m-r+1}, \ldots, x^m)$ *on M such that the metric tensor g takes the form:*

$$(g_{ij}) = \begin{pmatrix} B & H & \mathrm{Id}_r \\ {}^t H & A & 0 \\ \mathrm{Id}_r & 0 & 0 \end{pmatrix},$$

where Id_r is the identity matrix and A, B, and H are matrices whose entries are functions of $(x^1, \dots, x^{m-r}, x^{m-r+1}, \dots, x^m)$ such that A and B are symmetric matrices of order $m - 2r$ and r respectively, H is a $r \times (m - 2r)$ matrix and ${}^t H$ its transpose, and such that A and H are independent of the $(x^{m-r+1}, \dots, x^m)$ coordinates. Finally, the null parallel plane field $\mathfrak{D}$ is locally spanned by the coordinate vector fields $\{\partial_{x_{m-r+1}}, \dots, \partial_{x_m}\}$.

Remark 1.30   In the special case that $\dim(M) = 2\bar{m}$ and $\dim(\mathfrak{D}) = \frac{1}{2}\dim(M) = \bar{m}$ is maximal, there exist *Walker coordinates* $(x^1, \dots, x^{\bar{m}}, x_{1'}, \dots, x_{\bar{m}'})$ and there exists a symmetric $\bar{m} \times \bar{m}$ matrix $B = B(\vec{x}, \vec{x}')$ so that:

$$(g_{ij}) = \begin{pmatrix} B & \mathrm{Id}_{\bar{m}} \\ \mathrm{Id}_{\bar{m}} & 0 \end{pmatrix}.$$

The metrics given in Remark 1.30 are of special interest for our purposes. Their Christoffel symbols are given as follows (see Calviño-Louzao *et al.* [43]):

Lemma 1.31   *Let (M, g) be a $2\bar{m}$-dimensional Walker manifold with a null parallel plane field $\mathfrak{D}$ of dimension $\bar{m}$ and let $(x^1, \dots, x^{\bar{m}}, x_{1'}, \dots, x_{\bar{m}'})$ be adapted coordinates where the metric takes the form given in Remark 1.30. The (possibly) non-zero components of the Christoffel symbols of the Levi-Civita connection ${}^g\nabla$ are determined by:*

$$\Gamma_{ij}{}^k = -\tfrac{1}{2}\partial_{x^{k'}} g_{ij}, \quad \Gamma_{i'j}{}^{k'} = \tfrac{1}{2}\partial_{x^{i'}} g_{jk},$$

$$\Gamma_{ij}{}^{k'} = \tfrac{1}{2}\left(\partial_{x_j} g_{ik} - \partial_{x_k} g_{ij} + \partial_{x_i} g_{jk} + \sum_{1 \le s \le \bar{m}} g_{ks}\partial_{x_{s'}} g_{ij}\right).$$

Similarly, the (possibly) non-zero components of the curvature operator are determined by:

$$R_{jik}{}^h = \tfrac{1}{2}\left(\partial_{x_i}\partial_{x^{h'}} g_{jk} - \partial_{x_j}\partial_{x^{h'}} g_{ik}\right)$$

$$+ \tfrac{1}{4}\sum_{1 \le s \le \bar{m}}\left\{\partial_{x_{s'}} g_{ik}\partial_{x^{h'}} g_{js} - \partial_{x_{s'}} g_{jk}\partial_{x^{h'}} g_{is}\right\},$$

$$R_{jik}{}^{h'} = \tfrac{1}{2}\left(\partial_{x_j}\partial_{x_k} g_{ih} - \partial_{x_j}\partial_{x_h} g_{ik} + \partial_{x_i}\partial_{x_h} g_{jk} - \partial_{x_i}\partial_{x_k} g_{jh}\right)$$

$$+ \tfrac{1}{4}\sum_{1 \le s,t \le \bar{m}}\left\{\partial_{x_{s'}} g_{ik}\left(\partial_{x_h} g_{js} - \partial_{x_s} g_{jh} - \partial_{x_j} g_{sh} - g_{ht}\partial_{x_{t'}} g_{js}\right)\right.$$

$$- \partial_{x_{s'}} g_{jk}\left(\partial_{x_h} g_{is} - \partial_{x_s} g_{ih} - \partial_{x_i} g_{sh} - g_{ht}\partial_{x_{t'}} g_{is}\right)$$

$$- \partial_{x_{s'}} g_{jh}\left(\partial_{x_s} g_{ik} - \partial_{x_k} g_{is} - \partial_{x_i} g_{ks} - g_{st}\partial_{x_{t'}} g_{ik}\right)$$

$$+ \partial_{x_{s'}} g_{ih}\left(\partial_{x_s} g_{jk} - \partial_{x_k} g_{js} - \partial_{x_j} g_{ks} - g_{st}\partial_{x_{t'}} g_{jk}\right)$$

$$\left. + 2\partial_{x_j}\left(g_{hs}\partial_{x_{s'}} g_{ik}\right) - 2\partial_{x_i}\left(g_{hs}\partial_{x_{s'}} g_{jk}\right)\right\},$$

$$R_{ji'k}{}^h = \tfrac{1}{2}\partial_{x^{i'}}\partial_{x^{h'}} g_{jk},$$

$$R_{ji'k}{}^{h'} = \tfrac{1}{2}\left(\partial_{x_h}\partial_{x_{i'}}g_{jk} - \partial_{x_k}\partial_{x_{i'}}g_{jh}\right)$$
$$+ \tfrac{1}{4}\sum_s \left\{\partial_{x_{s'}}g_{jk}\partial_{x_{i'}}g_{sh} + \partial_{x_{s'}}g_{jh}\partial_{x_{i'}}g_{sk} - 2\partial_{x_{i'}}(g_{hs}\partial_{x_{s'}}g_{jk})\right\},$$
$$R_{jik'}{}^{h'} = \tfrac{1}{2}\left(\partial_{x_j}\partial_{x_{k'}}g_{ih} - \partial_{x_i}\partial_{x_{k'}}g_{jh}\right) + \tfrac{1}{4}\sum_s\left\{\partial_{x_{k'}}g_{is}\partial_{x_{s'}}g_{jh} - \partial_{x_{k'}}g_{js}\partial_{x_{s'}}g_{ih}\right\},$$
$$R_{ji'k'}{}^{h'} = -\tfrac{1}{2}\partial_{x_{i'}}\partial_{x_{k'}}g_{jh}.$$

1.7   METRICS ON THE COTANGENT BUNDLE

In this section, we continue the discussion of Section 1.6 and introduce special classes of Walker metrics by specializing those given in Remark 1.30. We shall discuss Riemannian extensions, deformed Riemannian extensions, and modified Riemannian extensions. We introduce these metrics in this section; many of their properties will be developed in subsequent chapters.

RIEMANNIAN EXTENSIONS

This class of manifolds was introduced by Patterson and Walker [177]. Their construction defines a Walker metric on the cotangent bundle T^*M of any affine manifold (M, ∇). The original construction of Patterson and Walker was later extended by Afifi [3]; this gives rise to a family of metrics on the cotangent bundle of any affine manifold. These metrics play a distinguished role in the study of some curvature problems. We adopt the notational conventions established in Section 1.1. Let ι be the evaluation map and let X^C be the complete lift of a vector field X on M. These are defined invariantly by the relations:

$$\iota X(P, \omega) = \ll X(P), \omega \gg \quad \text{and} \quad X^C(\iota Z) = \iota[X, Z].$$

In a system of local coordinates $(x^i, x_{i'})$ we have:

$$\iota X = X^i x_{i'} \quad \text{and} \quad X^C = X^j \partial_{x_j} - x_{i'}(\partial_{x_j}X^i)\partial_{x_{j'}} \quad \text{for } X = X^j \partial_{x_j}.$$

Definition 1.32   Let (M, ∇) be an affine manifold of dimension $\bar{m}$. The *Riemannian extension* g_∇ of (M, ∇) is the pseudo-Riemannian metric of neutral signature $(\bar{m}, \bar{m})$ on the cotangent bundle T^*M, which is characterized by the identity:

$$g_\nabla(X^C, Y^C) = -\iota(\nabla_X Y + \nabla_Y X).$$

In the system of induced coordinates $(x^i, x_{i'})$ on T^*M as described in Equation (1.1.d) and in Equation (1.1.e), the Riemannian extension takes the form:

$$g_\nabla = \begin{pmatrix} -2x_{k'}\Gamma_{ij}{}^k(\vec{x}) & \mathrm{Id}_{\bar{m}} \\ \mathrm{Id}_{\bar{m}} & 0 \end{pmatrix},$$

with respect to $\{\partial_{x_1}, \ldots, \partial_{x_{\bar{m}}}, \partial_{x^{1'}}, \ldots, \partial_{x^{\bar{m}'}}\}$; here the indices i and j range from $1, \ldots, \bar{m}$, $i' = i + \bar{m}$, and $\Gamma_{ij}{}^k$ are the Christoffel symbols of the connection ∇ with respect to the coordinates (x^i) on M. More explicitly:

$$g_\nabla(\partial_{x_i}, \partial_{x_j}) = -2x_{k'}\Gamma_{ij}{}^k(\vec{x}), \quad g_\nabla(\partial_{x_i}, \partial_{x_{j'}}) = \delta_i^j, \quad g_\nabla(\partial_{x_{i'}}, \partial_{x_{j'}}) = 0.$$

Let σ be the natural projection from T^*M to M. The Walker distribution is the null parallel plane field of maximal dimension $\bar{m}$ given by $\mathfrak{D} = \ker\{\sigma_*\}$.

Let (M, g) be a pseudo-Riemannian manifold. The Riemannian extension of the Levi-Civita connection inherits many of the properties of the base manifold. For instance, (M, g) has constant sectional curvature if and only if $(T^*M, g_{g\nabla})$ is locally conformally flat (see Cendán-Verdes, García-Río, and Vázquez-Abal [51]). However, the main applications of the Riemannian extensions appear when considering affine connections that are not the Levi-Civita connection of any metric. We refer to Yano and Ishihara [215] for the proof of the following result:

Lemma 1.33 *Let (M, ∇) be an affine manifold. Let $\tilde{\mathcal{R}}$ be the curvature operator of the Riemannian extension (T^*M, g_∇) and let $\mathcal{R}$ be the curvature operator of (M, ∇). Then:*

$$\tilde{R}_{kji}{}^h = R_{kji}{}^h, \quad \tilde{R}_{kji'}{}^{h'} = -R_{kjh}{}^i, \quad \tilde{R}_{kj'i}{}^{h'} = -R_{hik}{}^j, \quad \tilde{R}_{k'ji}{}^{h'} = -R_{hij}{}^k,$$

$$\tilde{R}_{kji}{}^{h'} = x_{a'}\left\{\nabla_{\partial_{x_h}} R_{kji}{}^a - \nabla_{\partial_{x_i}} R_{kjh}{}^a + \Gamma_{ht}{}^a R_{kji}{}^t + \Gamma_{kt}{}^a R_{ihj}{}^t\right.$$

$$\left. + \Gamma_{jt}{}^a R_{hik}{}^t + \Gamma_{it}{}^a R_{kjh}{}^t\right\},$$

where $R_{\alpha\beta\gamma}{}^\delta$ (resp. $\tilde{R}_{\alpha\beta\gamma}{}^\delta$) denote the components of $\mathcal{R}$ (resp. $\tilde{\mathcal{R}}$).

By Lemma 1.33, (T^*M, g_∇) is locally symmetric if and only if (M, ∇) is locally symmetric. Furthermore, (T^*M, g_∇) is locally conformally flat if and only if (M, ∇) is projectively flat (see Afifi [3]). Any projectively flat pseudo-Riemannian manifold has constant sectional curvature. However, there are many projectively flat affine connections. We refer to Weyl [208] and to Willmore [211] for further details.

DEFORMED RIEMANNIAN EXTENSIONS

The Riemannian extensions of Definition 1.32 can be generalized.

Definition 1.34 Let Φ be a symmetric $(0, 2)$-tensor field on an affine manifold (M, ∇) and let σ be the natural projection from T^*M to M. The *deformed Riemannian extension* $g_{\nabla, \Phi}$ is the metric of neutral signature $(\bar{m}, \bar{m})$ on the cotangent bundle given by:

$$g_{\nabla, \Phi} = g_\nabla + \sigma^*\Phi.$$

Let $\Gamma_{ij}{}^k$ be the Christoffel symbols of the connection ∇ and let Φ_{ij} be the local components of the symmetric $(0, 2)$-tensor field Φ. In local coordinates the deformed Riemannian extension is given by:

$$g_{\nabla,\Phi} = \begin{pmatrix} -2x_{k'}\Gamma_{ij}{}^k(\vec{x}) + \Phi_{ij}(\vec{x}) & \mathrm{Id}_{\bar{m}} \\ \mathrm{Id}_{\bar{m}} & 0 \end{pmatrix} \tag{1.7.a}$$

or, equivalently,

$$g_{\nabla,\Phi}(\partial_{x_i}, \partial_{x_j}) = -2x_{k'}\Gamma_{ij}{}^k(\vec{x}) + \Phi_{ij}(\vec{x}), \quad g_{\nabla,\Phi}(\partial_{x_i}, \partial_{x_{j'}}) = \delta_i^j, \quad g_{\nabla,\Phi}(\partial_{x_{i'}}, \partial_{x_{j'}}) = 0.$$

Note that the crucial terms $g_{\nabla,\Phi}(\partial_{x_i}, \partial_{x_j})$ now no longer vanish on the 0-section. As was the case for the Riemannian extension, the Walker distribution is the kernel of the projection from T^*M:

$$\mathfrak{D} = \ker\{\sigma_*\} = \mathrm{Span}\{\partial_{x_{i'}}\}.$$

The tensor Φ plays an essential role. Even if the underlying connection is flat, the deformed Riemannian extension need not be flat. Deformed Riemannian extensions are characterized by their curvature as follows (see Afifi [3]):

Lemma 1.35 *The Walker metric of Remark* 1.30 *is locally a deformed Riemannian extension if and only if the curvature tensor satisfies* $\mathcal{R}(\cdot, \mathfrak{D})\mathfrak{D} = 0$.

Proof. We use Lemma 1.31 to see that the coefficients g_{ij} of the metric are linear in the $x_{k'}$ variables. Thus, the metric locally takes the form given in Equation (1.7.a). □

Deformed Riemannian extensions have nilpotent Ricci operator and hence, they are Einstein if and only if they are Ricci flat. They can be used to construct non-flat Ricci flat pseudo-Riemannian manifolds:

Lemma 1.36 *A deformed Riemannian extension* $(T^*M, g_{\nabla,\Phi})$ *is Einstein if and only if it is Ricci flat. This happens if and only if the Ricci tensor of the connection* ∇ *is skew-symmetric.*

MODIFIED RIEMANNIAN EXTENSIONS

Definition 1.37 Let ∇ be a torsion-free connection and let Φ be a symmetric $(0, 2)$-tensor field on a manifold M. Let T, S be $(1, 1)$-tensor fields on M. The *modified Riemannian extension* is the metric on T^*M given by

$$g_{\nabla,\Phi,T,S} = \iota T \circ \iota S + g_{\nabla,\Phi}.$$

Let $(T \circ S)_{ij}^{rs} := \frac{1}{2}(T_i^r S_j^s + T_j^r S_i^s)$. In local coordinates:

$$g_{\nabla,\Phi,T,S} = \begin{pmatrix} x_{r'} x_{s'} (T \circ S)(\vec{x})_{ij}^{rs} - 2x_{k'} \Gamma_{ij}{}^k(\vec{x}) + \Phi_{ij}(\vec{x}) & \mathrm{Id}_{\bar{m}} \\ \mathrm{Id}_{\bar{m}} & 0 \end{pmatrix}. \tag{1.7.b}$$

In other words:

$$g(\partial_{x_i}, \partial_{x_j}) = x_{r'} x_{s'} (T \circ S)_{ij}^{rs}(\vec{x}) + \Phi_{ij}(\vec{x}) - 2x_{k'} \Gamma_{ij}{}^k(\vec{x}),$$
$$g(\partial_{x_i}, \partial_{x_{j'}}) = \delta_i^j, \quad \text{and} \quad g(\partial_{x_{i'}}, \partial_{x_{j'}}) = 0.$$

The particular case where $T = c\,\mathrm{Id}$ and $S = \mathrm{Id}$ will be denoted by $g_{\nabla,\Phi,c}$. The crucial point is that this is now *quadratic* in the fiber coordinates $x_{i'}$.

As well as in the previous cases, modified Riemannian extensions are Walker metrics where the Walker distribution is given by $\mathfrak{D} = \ker\{\sigma_*\}$. Modified Riemannian extensions are characterized by their curvature (see Afifi [3]). In particular, any locally symmetric Walker metric, as given in Remark 1.30, is locally a modified Riemannian extension.

Lemma 1.38   *The Walker metric of Remark 1.30 is locally a modified Riemannian extension if and only if the covariant derivative of the curvature operator satisfies $(\nabla_{\mathfrak{D}} R)(\cdot, \mathfrak{D})\mathfrak{D} = 0$.*

The following is a very general observation and gives rise to a wide family of neutral signature Einstein metrics which are not Ricci flat:

Lemma 1.39   *Let (M, ∇) be an affine manifold of dimension $\bar{m}$ and let $c \neq 0$. The modified Riemannian extension $(T^*M, g_{\nabla,\Phi,c})$ is Einstein if and only if $\Phi = \frac{4}{c(\bar{m}-1)}\rho_s$.*

Proof. Let $\tau = 6c$ be the scalar curvature of the modified Riemannian extension $g = g_{\nabla,\Phi,c}$. We use Lemma 1.31 to see that the trace-free Ricci tensor

$$g_{\rho_0} = {}^g\rho - \frac{\tau}{2\bar{m}} g$$

of g is given by

$$g_{\rho_0} = 2\sigma^* \rho_s - \frac{1}{2}c(\bar{m} - 1)\sigma^*\Phi. \qquad \square$$

1.8   SELF-DUAL WALKER METRICS

Four-dimensional geometry is exceptional in many respects. In this section, we discuss the Hodge $\star$ operator and self-duality. Throughout this section, let $(V, \langle \cdot, \cdot \rangle)$ be an oriented inner product space of signature (p, q). Let $\{e_1, e_2, e_3, e_4\}$ be an orthonormal basis so that the orientation is given by

$$e^1 \wedge e^2 \wedge e^3 \wedge e^4.$$

THE HODGE ⋆ OPERATOR

Let $\star$ be the *Hodge operator*. This operator is characterized by the relation:

$$\omega_1 \wedge \star\omega_2 = \langle \omega_1, \omega_2 \rangle e^1 \wedge e^2 \wedge e^3 \wedge e^4 \text{ for all } \omega_i \in \Lambda^2(V^*),$$

where we denote by $\langle \omega_1, \omega_2 \rangle$ the induced product on $\Lambda^2 = \Lambda^2(V^*)$. Let $\varepsilon_i = \langle e_i, e_i \rangle = \pm 1$. Then:

$$e^i \wedge e^j \wedge \star(e^k \wedge e^\ell) = (\delta_k^i \delta_\ell^j - \delta_\ell^i \delta_k^j)\, \varepsilon_i \varepsilon_j\, e^1 \wedge e^2 \wedge e^3 \wedge e^4.$$

Since $\star^2 = \mathrm{Id}$ if $\langle \cdot, \cdot \rangle$ is definite or has neutral signature, the Hodge $\star$ operator induces a splitting $\Lambda^2 = \Lambda^+ \oplus \Lambda^-$, where Λ^+ and Λ^- denote the spaces of self-dual and anti-self-dual 2-forms

$$\Lambda^\pm = \{\alpha \in \Lambda^2 : \star\alpha = \pm\alpha\}.$$

Furthermore, the induced inner products on $\Lambda^\pm$ are positive definite if $\langle \cdot, \cdot \rangle$ is Riemannian, but they are Lorentzian if $\langle \cdot, \cdot \rangle$ is of neutral signature. An orthonormal basis for the self-dual and anti-self-dual space is given by

$$\Lambda^\pm = \mathrm{Span}\left\{ E_1^\pm = \frac{e^1 \wedge e^2 \pm \varepsilon_3\varepsilon_4 e^3 \wedge e^4}{\sqrt{2}},\ E_2^\pm = \frac{e^1 \wedge e^3 \mp \varepsilon_2\varepsilon_4 e^2 \wedge e^4}{\sqrt{2}}, \right.$$
$$\left. E_3^\pm = \frac{e^1 \wedge e^4 \pm \varepsilon_2\varepsilon_3 e^2 \wedge e^3}{\sqrt{2}} \right\}. \tag{1.8.a}$$

Henceforth, we will use these two bases when coordinates in the self-dual or anti-self-dual space are needed, except where indicated explicitly to the contrary. The Weyl operator in the Singer–Thorpe decomposition of Theorem 1.24 further splits in the form:

$$\mathcal{W} = \mathcal{W}^+ + \mathcal{W}^- \text{ where } \mathcal{W}^\pm := \tfrac{1}{2}\left(\mathcal{W} \pm \star\mathcal{W}\right).$$

The curvature operator is said to be *self-dual* (resp. *anti-self-dual*) if we have $\mathcal{W}^- = 0$ (resp. $\mathcal{W}^+ = 0$).

We now pass to the geometric setting. We adopt the notation of Equation (1.1.b) and let δ be the *coderivative* mapping $C^\infty(\Lambda^p M)$ to $C^\infty(\Lambda^{p-1} M)$; δ is the L^2 adjoint of d. We can express δ in terms of the Hodge $\star$ operator:

$$\delta = (-1)^{mp+m+1} \star d\star = (-1)^p \star^{-1} d\star. \tag{1.8.b}$$

Let $m = 4$. If (M, g, J_+) is an almost para-Hermitian manifold of signature $(2, 2)$, or if (M, g, J_-) is a Hermitian manifold of positive definite signature $(0, 4)$, then $J_\pm$ defines an orientation of M so that $\Omega_\pm$ is a self-dual 2-form. However, if (M, g, J_-) is a 4-dimensional almost Hermitian manifold with a metric of neutral signature $(2, 2)$, then Ω_- is anti-self-dual.

The special significance of the (anti-)self-dual condition is that a 4-dimensional almost (para)-Hermitian structure is self-dual if and only if the Bochner curvature tensor described in Remark 1.26 vanishes (see, for example, Bryant [39] or Tricerri and Vanhecke [200]). In the almost pseudo-Hermitian case of signature $(2, 2)$, the Bochner tensor vanishes if and only if the metric is anti-self-dual.

SELF-DUALITY

The action of the Hodge $\star$ operator on the space of self-dual Walker metrics was investigated by Díaz-Ramos, García-Río, and Vázquez-Lorenzo [71]; a 4-dimensional Walker metric is self-dual if and only the metric of Remark 1.30 in Walker coordinates $(x^1, x^2, x_{1'}, x_{2'})$ takes the form:

$$
\begin{aligned}
g_{11}(x^1, x^2, x_{1'}, x_{2'}) &= x_{1'}^3 \mathcal{A} + x_{1'}^2 \mathcal{B} + x_{1'}^2 x_{2'} \mathcal{C} + x_{1'} x_{2'} \mathcal{D} + x_{1'} P + x_{2'} Q + \xi, \\
g_{22}(x^1, x^2, x_{1'}, x_{2'}) &= x_{2'}^3 \mathcal{C} + x_{2'}^2 \mathcal{E} + x_{1'} x_{2'}^2 \mathcal{A} + x_{1'} x_{2'} \mathcal{F} + x_{1'} S + x_{2'} T + \eta, \\
g_{12}(x^1, x^2, x_{1'}, x_{2'}) &= \tfrac{1}{2} x_{1'}^2 \mathcal{F} + \tfrac{1}{2} x_{2'}^2 \mathcal{D} + x_{1'}^2 x_{2'} \mathcal{A} + x_{1'} x_{2'}^2 \mathcal{C} + \tfrac{1}{2} x_{1'} x_{2'} (\mathcal{B} + \mathcal{E}) \\
&\quad + x_{1'} U + x_{2'} V + \gamma,
\end{aligned}
\tag{1.8.c}
$$

where the coefficients are smooth functions of (x^1, x^2). It was observed in Calviño-Louzao *et al.* [43] that the metrics of Equation (1.8.c) can be realized on the cotangent bundle of an affine surface Σ by means of a slight generalization of the modified Riemannian extensions as follows.

Theorem 1.40   *A 4-dimensional Walker manifold is self-dual if and only if it is locally isometric to the cotangent bundle $T^*\Sigma$ of an affine surface (Σ, ∇), with a metric tensor*

$$
g = \iota X (\iota \, \mathrm{Id} \circ \iota \, \mathrm{Id}) + \iota \, \mathrm{Id} \circ \iota T + g_\nabla + \sigma^* \Phi ,
$$

where X, T, ∇ and Φ are a vector field, a $(1, 1)$-tensor field, a torsion-free affine connection, and a symmetric $(0, 2)$-tensor field on Σ, respectively.

Proof. Set $X = \mathcal{A}(x^1, x^2)\partial_{x_1} + \mathcal{C}(x^1, x^2)\partial_{x_2}$ to be a locally defined vector field on Σ so that it defines a function ιX on $T^*\Sigma$ as follows:

$$
\iota X = x_{1'} \mathcal{A}(x^1, x^2) + x_{2'} \mathcal{C}(x^1, x^2).
$$

Hence,

$$
\begin{aligned}
(\iota X \cdot \iota \, \mathrm{Id} \circ \iota \, \mathrm{Id})_{11} &= x_{1'}^3 \mathcal{A}(x^1, x^2) + x_{1'}^2 x_{2'} \mathcal{C}(x^1, x^2), \\
(\iota X \cdot \iota \, \mathrm{Id} \circ \iota \, \mathrm{Id})_{12} &= x_{1'}^2 x_{2'} \mathcal{A}(x^1, x^2) + x_{1'} x_{2'}^2 \mathcal{C}(x^1, x^2), \\
(\iota X \cdot \iota \, \mathrm{Id} \circ \iota \, \mathrm{Id})_{22} &= x_{1'} x_{2'}^2 \mathcal{A}(x^1, x^2) + x_{2'}^3 \mathcal{C}(x^1, x^2).
\end{aligned}
$$

Next, define locally a $(1, 1)$-tensor field T on Σ by specializing its components by

$$
T_1^1 = \mathcal{B}(x^1, x^2), \quad T_1^2 = \mathcal{D}(x^1, x^2), \quad T_2^1 = \mathcal{F}(x^1, x^2), \quad T_2^2 = \mathcal{E}(x^1, x^2).
$$

Now the action of ι on T described in Section 1.1 gives a 1-form ιT on $T^*\Sigma$ with components

$$
\begin{aligned}
(\iota T)_1 &= x_{1'} \mathcal{B}(x^1, x^2) + x_{2'} \mathcal{D}(x^1, x^2), \\
(\iota T)_2 &= x_{1'} \mathcal{F}(x^1, x^2) + x_{2'} \mathcal{E}(x^1, x^2),
\end{aligned}
$$

and therefore,

$$(\iota T \circ \iota \operatorname{Id})_{11} = x_{1'}^2 \mathcal{B}(x^1, x^2) + x_{1'} x_{2'} \mathcal{D}(x^1, x^2),$$
$$(\iota T \circ \iota \operatorname{Id})_{12} = \tfrac{1}{2} \left(x_{1'}^2 \mathcal{F}(x^1, x^2) + x_{2'}^2 \mathcal{D}(x^1, x^2) \right.$$
$$\left. + x_{1'} x_{2'} (\mathcal{B}(x^1, x^2) + \mathcal{E}(x^1, x^2)) \right),$$
$$(\iota T \circ \iota \operatorname{Id})_{22} = x_{1'} x_{2'} \mathcal{F}(x^1, x^2) + x_{2'}^2 \mathcal{E}(x^1, x^2).$$

Finally, consider the affine connection ∇ locally defined by the Christoffel symbols

$$\Gamma_{11}{}^1 = -\tfrac{1}{2} P(x^1, x^2), \quad \Gamma_{12}{}^1 = -\tfrac{1}{2} S(x^1, x^2), \quad \Gamma_{22}{}^1 = -\tfrac{1}{2} U(x^1, x^2),$$
$$\Gamma_{11}{}^2 = -\tfrac{1}{2} Q(x^1, x^2), \quad \Gamma_{12}{}^2 = -\tfrac{1}{2} T(x^1, x^2), \quad \Gamma_{22}{}^2 = -\tfrac{1}{2} V(x^1, x^2),$$

and the symmetric $(0, 2)$-tensor field Φ locally given by

$$\Phi_{11} = \xi(x^1, x^2), \quad \Phi_{12} = \eta(x^1, x^2), \quad \Phi_{22} = \gamma(x^1, x^2).$$

Now the proof follows directly from Equation (1.8.c). □

Remark 1.41   Any 4-dimensional para-Kähler metric is necessarily a Walker metric. Consequently, the distributions corresponding to the ± 1-eigenspaces of the para-complex structure J_+ form a null parallel plane field of maximal dimension. The underlying structure of any 4-dimensional Bochner flat para-Kähler metric is given in Theorem 1.40.

1.9   RECURRENT CURVATURE

An affine connection ∇ is said to be *recurrent* (or to have *recurrent curvature*) if there exists a 1-form ω such that the covariant derivative of the curvature operator $\mathcal{R}$ satisfies $\nabla \mathcal{R} = \omega \otimes \mathcal{R}$ i.e.,

$$\nabla \mathcal{R}(X_1, X_2; X_4) X_3 = \omega(X_4) \mathcal{R}(X_1, X_2) X_3.$$

As the curvature of an affine surface (Σ, ∇) satisfies (see, for example, Nomizu [166])

$$\mathcal{R}(X, Y) Z = \rho(Y, Z) X - \rho(X, Z) Y,$$

the curvature operator is recurrent if and only if the Ricci tensor is recurrent. Recurrent affine surfaces have been classified by Wong [213]. The classification depends on the different possibilities for the symmetric and the alternating Ricci tensors. Affine surfaces with recurrent curvature play an important role in understanding the geometry of Riemannian extensions as we shall see presently. The following result is based on work of Derdzinski [66] and of Wong [213].

Theorem 1.42   *Let (Σ, ∇) be an affine surface with recurrent curvature. There exists a coordinate system (x^1, x^2) where the (possibly) non-zero Christoffel symbols are given by:*

1. *If* $\mathrm{Rank}\{\rho_s\} = 1$ *and* $\rho_a \neq 0$, *then* $\Gamma_{11}{}^1 = \Gamma_{12}{}^2 = \partial_{x_2}\theta$ *and* $\Gamma_{11}{}^2 = \partial_{x_1}\theta - \partial_{x_2}\theta$ *where* $\theta = \theta(x^1, x^2)$ *satisfies* $\partial_{x_2}\partial_{x_2}\theta \neq 0$.

2. *If* $\mathrm{Rank}\{\rho_s\} = 1$ *and* $\rho_a = 0$, *then the only non-zero Christoffel symbol takes the form* $\Gamma_{11}{}^2 = \Gamma_{11}{}^2(x^1, x^2)$ *where* $\partial_{x_2}\Gamma_{11}{}^2 \neq 0$.

3. *If* $\det\{\rho_s\} < 0$ *and* $\rho_a \neq 0$, *then either the only non-zero Christoffel symbol is* $\Gamma_{22}{}^2$ *with* $\partial_{x_1}\Gamma_{22}{}^2 \neq 0$, *or* $\Gamma_{11}{}^1 = (1-c)^{-1}\partial_{x_1}\theta$ *and* $\Gamma_{22}{}^2 = (1+c)^{-1}\partial_{x_2}\theta$ *where* $\theta = \theta(x^1, x^2)$ *satisfies* $\partial_{x_1}\partial_{x_2}\theta \neq 0$ *and where* $c \neq 0, \pm 1$.

4. *If* $\det\{\rho_s\} < 0$ *and* $\rho_a = 0$, *then* $\Gamma_{11}{}^1 = \partial_{x_1}\theta$ *and* $\Gamma_{22}{}^2 = \partial_{x_2}\theta$ *where* $\theta = \theta(x^1, x^2)$ *satisfies* $\partial_{x_1}\partial_{x_2}\theta \neq 0$.

5. *If* $\det\{\rho_s\} > 0$ *and* $\rho_a \neq 0$, *then there exists* $\theta = \theta(x^1, x^2)$ *with* $\partial_{x_1}\partial_{x_1}\theta + \partial_{x_2}\partial_{x_2}\theta \neq 0$ *and there exists* $0 \neq c \in \mathbb{R}$ *so that*

$$\Gamma_{11}{}^1 = \Gamma_{12}{}^2 = -\Gamma_{22}{}^1 = (1+c^2)^{-1}\left(\partial_{x_1}\theta + c\partial_{x_2}\theta\right),$$
$$\Gamma_{22}{}^2 = \Gamma_{12}{}^1 = -\Gamma_{11}{}^2 = (1+c^2)^{-1}\left(\partial_{x_2}\theta - c\partial_{x_1}\theta\right).$$

6. *If* $\det\{\rho_s\} > 0$ *and* $\rho_a = 0$, *then there exists* $\theta = \theta(x^1, x^2)$ *with* $\partial_{x_1}\partial_{x_1}\theta + \partial_{x_2}\partial_{x_2}\theta \neq 0$ *so that* $\Gamma_{11}{}^1 = \Gamma_{12}{}^2 = -\Gamma_{22}{}^1 = \partial_{x_1}\theta$ *and* $\Gamma_{22}{}^2 = \Gamma_{12}{}^1 = -\Gamma_{11}{}^2 = \partial_{x_2}\theta$.

7. *If* $\rho_s = 0$ *and* $\rho_a \neq 0$, *then there exists* $\theta = \theta(x^1, x^2)$ *with* $\partial_{x_1}\partial_{x_2}\theta \neq 0$ *so that we have* $\Gamma_{11}{}^1 = -\partial_{x_1}\theta$ *and* $\Gamma_{22}{}^2 = \partial_{x_2}\theta$.

Remark 1.43 The Ricci tensor is symmetric and non-degenerate in Cases (4) and (6). Furthermore, in these cases, there is a function f locally defined on Σ such that $g = e^f \rho$ is a metric tensor on Σ such that ∇ is its Levi-Civita connection (see Wong [213] for details).

1.10 CONSTANT CURVATURE

In this section, we shall discuss the Jacobi operator, symmetric spaces, manifolds of constant sectional curvature, and Kähler manifolds of constant holomorphic sectional curvature. A vector field X along a geodesic γ is said to be a Jacobi vector field if

$$\ddot{X} + \mathcal{R}(X, \dot{\gamma})\dot{\gamma} = 0.$$

We use Jacobi vector fields to show that the geodesic involution is an isometry if and only if $\nabla\mathcal{R} = 0$. We also use Jacobi vector fields to show that the local geometry of a pseudo-Riemannian manifold of signature (p, q) and of constant sectional curvature c is determined by (p, q, c). We conclude the section with a brief introduction to the theory of manifolds of constant holomorphic sectional curvature.

THE JACOBI OPERATOR AND JACOBI VECTOR FIELDS

The *Jacobi operator* $\mathcal{J}(X)$ is an endomorphism of the tangent bundle, which is defined for a tangent vector X by setting

$$\mathcal{J}(X) : Y \to \mathcal{R}(Y, X)X \text{ for } Y \in C^\infty(TM) . \tag{1.10.a}$$

It plays an important role in the study of geodesic sprays, which we outline as follows. Let ∇ be an affine connection and let $\gamma(s)$ be a geodesic in M. Recall that a vector field X along γ is said to be a *Jacobi vector field* if it satisfies the equation

$$\ddot{X} + \mathcal{J}(\dot{\gamma})X = 0 \text{ i.e., } \nabla_{\dot{\gamma}} \nabla_{\dot{\gamma}} X + \mathcal{R}(X, \dot{\gamma})\dot{\gamma} = 0 .$$

One says that a smooth map $T : [0, \varepsilon] \times [0, \varepsilon] \to M$ is a *geodesic spray* if T is an embedding such that the curves $\gamma_t(s) := T(t, s)$ are geodesics for all t. One has (see, for example, do Carmo [74]):

Lemma 1.44 *Let $T(t, s)$ be a geodesic spray. Then the variation $T_*(\partial_t)$ is a Jacobi vector field along the geodesics $\gamma_t(s) = T(t, s)$.*

Proof. To simplify the notation, we identify ∂_t with $\gamma_* \partial_t$ and ∂_s with $\gamma_* \partial_s$. Since the curves $s \to \gamma_t(s)$ are geodesics for all t, we may compute:

$$0 = \nabla_{\partial_t} \nabla_{\partial_s} \partial_s \qquad \text{(The curves } \gamma_t(s) \text{ are geodesics for any } t)$$

$$= \mathcal{R}(\partial_t, \partial_s)\partial_s + \nabla_{\partial_s} \nabla_{\partial_t} \partial_s \qquad \text{(Definition of the curvature; } [\partial_t, \partial_s] = 0)$$

$$= \mathcal{J}(\partial_s)\partial_t + \nabla_{\partial_s} \nabla_{\partial_s} \partial_t \qquad \text{(}\nabla \text{ is a torsion-free connection)}$$

$$= \mathcal{J}(\dot{\gamma})X + \ddot{X} . \qquad \text{(Disentangling the notation)}$$

The desired result now follows. $\square$

SYMMETRIC SPACES

Jacobi vector fields appear in many contexts. We present an example, arising from work of Cartan [49, 50], to illustrate their use and to motivate the study of the spectral geometry of the Jacobi operator in Section 1.11. Let (M, g) be a pseudo-Riemannian manifold. Let $\exp_P$ be the exponential map; this is the local diffeomorphism from a neighborhood of 0 in $T_P M$ to a neighborhood of P in M, defined in Section 1.2; it is characterized by the fact that the curves $s \to \exp_P(sv)$ are geodesics starting at P with initial direction v for $v \in T_P M$. The *geodesic symmetry at P* is then defined on a suitable neighborhood of P by setting:

$$S_P(Q) = \exp_P\{- \exp_P^{-1} Q\} .$$

We may use $\exp_P$ to identify a neighborhood of 0 in $T_P M$ with a neighborhood of P in M and to regard $T_P M$ as a pseudo-Riemannian manifold locally isometric to M; under this identification,

$S_P(v) = -v$ is simply multiplication by -1 and the straight lines through the origin are geodesics. One has the following result:

Lemma 1.45   *Let (M, g) be a connected pseudo-Riemannian manifold.*

1. *The following assertions are equivalent and, if either is satisfied, then (M, g) is said to be* locally symmetric *or to be a* local symmetric space*:*

   (a) *The geodesic symmetry S_P is an isometry for all $P \in M$.*

   (b) *$\nabla \mathcal{R} = 0$.*

2. *If (M, g) is locally symmetric, then (M, g) is* locally homogeneous, *i.e., given any two points $P, Q \in M$, there is an isometry $\Theta_{P,Q}$ from some neighborhood of P in M to some neighborhood of Q in M.*

Remark 1.46   If (M, g) is a local symmetric space that is complete and simply connected, then (M, g) is said to be *globally symmetric* or to be a *symmetric space*. In this setting, the geodesic symmetry extends to a global isometry of (M, g), and (M, g) is homogeneous. If G_0 is the isotropy subgroup of the group of isometries G of (M, g), then $M = G/G_0$ with the induced metric. Global symmetric spaces form a very special class of pseudo-Riemannian manifolds and techniques of group theory are used to study them—the associated Lie algebras $\mathfrak{g}$ and $\mathfrak{g}_0$ of G and of G_0 play a central role. We refer to Helgason [123] for further details.

Proof. We use $\exp_P$ to identify a neighborhood of 0 in $T_P M$ with a neighborhood of P in M henceforth. Suppose first, that S_P is an isometry. Then S_P commutes with ∇. Thus,

$$\nabla \mathcal{R}_P(-X, -Y; -Z)(-T) = -\nabla \mathcal{R}_P(X, Y; Z)T .$$

This implies that $\nabla \mathcal{R}_P = 0$ and shows that Assertion (1a) implies Assertion (1b).

We use Jacobi vector fields to show that the reverse implication holds. Suppose Assertion (1b) holds. Fix $v \in T_P M$ and consider the geodesic spray $T(t, s; w) := s(v + tw)$. The associated Jacobi vector field of Lemma 1.44 is given by $Y(s) = sw$; Y is determined by the Jacobi equation:

$$\ddot{Y} + J(\dot{\gamma})Y = 0 \text{ with } Y(0) = 0 \text{ and } \dot{Y}(0) = w .$$

Let $\{e_i\}$ be a parallel orthonormal frame along $\dot{\gamma}$. Expand

$$\{\mathcal{R}(e_i, \dot{\gamma})\dot{\gamma}\}(s) = c_i^j(s)e_j(s) .$$

We compute:

$$\begin{aligned} \nabla_{\dot{\gamma}}\{c_i^j e_j\} &= \dot{c}_i^j \cdot e_j = \nabla_{\dot{\gamma}}\{\mathcal{R}(e_i, \dot{\gamma})\dot{\gamma}\} \\ &= (\nabla_{\dot{\gamma}}\mathcal{R})(e_i, \dot{\gamma})\dot{\gamma} + \mathcal{R}(\nabla_{\dot{\gamma}}e_i, \dot{\gamma})\dot{\gamma} + \mathcal{R}(e_i, \nabla_{\dot{\gamma}}\dot{\gamma})\dot{\gamma} + \mathcal{R}(e_i, \dot{\gamma})(\nabla_{\dot{\gamma}}\dot{\gamma}) = 0. \end{aligned}$$

Consequently, the structure functions c_i^j are in fact constant. Let $Y(s) = sw = a^j(s)e_j(s)$. Then the Jacobi equation becomes

$$\ddot{a}^j(s) + a^k(s)c_k^i = 0 \text{ for } 1 \leq j \leq m \text{ with } a^j(0) = 0 \text{ and } \dot{a}^j(0) = g_0(w, e_j). \qquad (1.10.\text{b})$$

Let $b^j(s) = -a^j(-s)$. The $b^j(s)$ also satisfy Equation (1.10.b) and, hence, by the uniqueness of the solution to an ordinary differential equation, we have $b^j(s) = a^j(s)$, i.e., we have that $a^j(-s) = -a^j(s)$. We have $g_{\gamma(s)}(e_i(s), e_j(s)) = g_0(e_i, e_j)$ is independent of s. Consequently,

$$\begin{aligned} g_\gamma(s)(Y(s), Y(s)) &= a^i(s)a^j(s)g_{\gamma(s)}(e_i(s), e_j(s)) \\ &= a^i(-s)a^j(-s)g_{\gamma(-s)}(e_i(-s), e_j(-s)) = g_{\gamma(-s)}(Y(-s), Y(-s)). \end{aligned}$$

This implies

$$s^2 g_{sv}(w, w) = g_{\gamma(s)}(Y(s), Y(s)) = g_{\gamma(-s)}(Y(-s), Y(-s)) = s^2 g_{-sv}(w, w).$$

This implies $g_u = g_{-u}$ for $u \neq 0$ and thus, S_P is an isometry on a punctured neighborhood of P; continuity then implies S_P is an isometry at P, as well. Thus, Assertion (1b) implies Assertion (1a); this completes the proof of Assertion (1).

Assume that the geodesic involution is always an isometry. We establish Assertion (2) as follows. Suppose there is a geodesic γ from P to Q. We may use the geodesic involution around the midpoint of γ to construct a local isometry interchanging P and Q. As M is connected, we can compose a succession of such isometries to construct a local isometry between any two points of M and thereby show (M, g) is locally homogeneous. □

MANIFOLDS OF CONSTANT SECTIONAL CURVATURE

Previously, we used Jacobi vector fields to show the equivalence of a geometric condition (geodesic symmetries are isometries) with a purely algebraic condition (the curvature is parallel). In this section, we again use Jacobi vector fields to establish the equivalence of a geometric condition with an algebraic one. Let π be a non-degenerate 2-plane in $T_P M$. We define the *sectional curvature* $\kappa(\pi)$ by setting:

$$\kappa(\pi) := \frac{R(x, y, y, x)}{g(x, x)g(y, y) - g(x, y)^2} \text{ for any basis } \{x, y\} \text{ for } \pi.$$

We say that a pseudo-Riemannian manifold (M, g) has *constant sectional curvature* c if $\kappa(\pi) = c$ for any non-degenerate 2-plane π at any point P of M. The requirement that π be non-degenerate is, of course, equivalent to the condition $g(x, x)g(y, y) - g(x, y)^2 \neq 0$; this condition is independent of the particular basis chosen.

We now give a purely algebraic formalism. Recall that a *Riemannian curvature model* is a triple $(V, \langle \cdot, \cdot \rangle, A)$, where $(V, \langle \cdot, \cdot \rangle)$ is an inner product space and where $A \in \otimes^4 V^*$ satisfies the identities of the Riemann curvature tensor given in Equations (1.2.l)-(1.2.n).

Lemma 1.47   *Let $\mathfrak{M} := (V, \langle \cdot, \cdot \rangle, A)$ be a Riemannian curvature model.*

1. *The following conditions are equivalent and, if either is satisfied, then $\mathfrak{M}$ is said to have* constant sectional curvature c.

   (a) $A(x, y, z, w) = c\{\langle x, w \rangle \langle y, z \rangle - \langle x, z \rangle \langle y, w \rangle\}$.

   (b) $\dfrac{A(e_1, e_2, e_2, e_1)}{\langle e_1, e_1 \rangle \langle e_2, e_2 \rangle - \langle e_1, e_2 \rangle^2} = c$ *for any non-degenerate 2-plane* $\pi = \mathrm{Span}\{e_1, e_2\}$ *in* V.

2. *If $\mathfrak{M}$ has constant sectional curvature c and if $\{e_1, e_3\}$ is an orthonormal set, then we have that*
$$\mathcal{J}(e_1)e_3 = c\langle e_1, e_1 \rangle e_3.$$

Remark 1.48   The sectional curvature is a continuous function on the Grassmannian $\mathrm{Gr}_2^0(V)$ of non-degenerate 2-planes in V. In the positive definite setting, this space is compact and consequently the sectional curvature is bounded. This fact does not hold true in the indefinite setting since $\mathrm{Gr}_2^0(V)$ is an open subset of the full Grassmannian $\mathrm{Gr}_2(V)$ and, in fact, the sectional curvature is bounded if and only if it is constant. Furthermore, $R(x, y, y, x) = 0$ whenever x, y span a degenerate plane if and only if the sectional curvature is constant. We refer to O'Neill [169] for more information.

Proof. It is immediate that Assertion (1a) implies Assertion (1b). Assume, conversely, that Assertion (1b) holds. Let $\{x, y, z\}$ be an orthonormal set of vectors in V. Set:

$$\varepsilon = \langle x, x \rangle \langle y, y \rangle = \pm 1,$$
$$\xi(t) := \left\{ \begin{array}{ll} \cos(t)x + \sin(t)y & \text{if } \varepsilon = 1 \\ \cosh(t)x + \sinh(t)y & \text{if } \varepsilon = -1 \end{array} \right\}.$$

We may then compute that

$$\langle \xi(t), \xi(t) \rangle = \langle x, x \rangle \cdot \left\{ \begin{array}{ll} \cos^2(t) + \sin^2(t) & \text{if } \varepsilon = 1 \\ \cosh^2(t) - \sinh^2(t) & \text{if } \varepsilon = -1 \end{array} \right\} = \langle x, x \rangle = \pm 1$$

is independent of the parameter t. As $\mathfrak{M}$ has constant sectional curvature c, we have that:

$$\begin{aligned} c\langle x, x \rangle \langle z, z \rangle &= c\langle \xi(t), \xi(t) \rangle \langle z, z \rangle = A(\xi(t), z, z, \xi(t)) \\ &= c \left\{ \begin{array}{ll} \cos^2(t) + \sin^2(t) & \text{if } \varepsilon = 1 \\ \cosh^2(t) - \sinh^2(t) & \text{if } \varepsilon = -1 \end{array} \right\} \langle x, x \rangle \langle z, z \rangle \\ &\quad + 2A(x, z, z, y) \left\{ \begin{array}{ll} \cos(t)\sin(t) & \text{if } \varepsilon = 1 \\ \cosh(t)\sinh(t) & \text{if } \varepsilon = -1 \end{array} \right\}. \end{aligned}$$

This shows that:

$$A(x, z, z, y) = 0 \text{ if } \{x, y, z\} \text{ is an orthonormal set}. \tag{1.10.c}$$

Suppose $\{x, y, z, w\}$ is an orthonormal set. We argue similarly and polarize the identity of Equation (1.10.c) to conclude

$$A(x, z, w, y) + A(x, w, z, y) = 0. \tag{1.10.d}$$

We now use the curvature symmetries together with Equation (1.10.d) to see:

$$
\begin{aligned}
0 &= A(x, y, z, w) + A(y, z, x, w) + A(z, x, y, w) \\
&= A(x, y, z, w) - A(y, x, z, w) - A(x, z, y, w) \\
&= 3A(x, y, z, w) \text{ so}
\end{aligned}
$$

$$A(x, y, z, w) = 0 \text{ if } \{x, y, z, w\} \text{ is an orthonormal set}. \tag{1.10.e}$$

We verify that Assertion (1b) implies Assertion (1a) as follows. Let $\{e_1, \ldots, e_m\}$ be an orthonormal basis for V. By the curvature symmetries, $A(e_i, e_j, e_k, e_\ell) = 0$ unless $\{i, j\}$ and $\{k, \ell\}$ are distinct. By Equation (1.10.c) and Equation (1.10.e), $A(e_i, e_j, e_k, e_\ell) = 0$ unless either $(i, j) = (k, \ell)$ or $(i, \ell) = (j, k)$. The desired result now follows from the identity of Assertion (1b).

Let $\mathfrak{M}$ have constant sectional curvature c. Let $\{e_1, e_2, e_3\}$ be an orthonormal set of vectors in V; Assertion (1a) implies that $A(e_3, e_1, e_1, e_2) = 0$ so $\mathcal{J}(e_1)e_3 \perp e_2$. Since $A(e_3, e_1, e_1, e_1) = 0$, we conclude that $\mathcal{J}(e_1)e_3$ is perpendicular to e_1 and e_2. Since e_2 was an arbitrary unit vector perpendicular to e_1 and e_3, we conclude $\mathcal{J}(e_1)e_3$ is some multiple λ of e_3. We establish Assertion (2) by computing:

$$c\langle e_1, e_1\rangle\langle e_3, e_3\rangle = A(e_3, e_1, e_1, e_3) = \langle \mathcal{J}(e_1)e_3, e_3\rangle = \lambda\langle e_3, e_3\rangle. \qquad \square$$

Let (M, g) be a pseudo-Riemannian manifold. If $(T_P M, g_P, R_P)$ has constant sectional curvature c_P at each point P of M and if $m \geq 3$, then a Schur type lemma shows that, in fact, c_P is actually constant. Such manifolds are often also called *space forms*; if $c > 0$, the manifold is said to be a *spherical space form* while if $c < 0$, it is said to be a *hyperbolic space form*. If $c = 0$, then the manifold is *flat* since the curvature tensor vanishes identically. We refer to Wolf [212] for further information and content ourselves with using Lemma 1.47 to establish the following result:

Lemma 1.49 *Let (M, g) and $(\tilde{M}, \tilde{g})$ be two connected pseudo-Riemannian manifolds of constant sectional curvature c and signature (p, q). Let P and $\tilde{P}$ be points of M and $\tilde{M}$. Then there is a local isometry Θ from M to $\tilde{M}$ with $\Theta(P) = \tilde{P}$.*

Proof. We use $\exp_P$ to identify a neighborhood of 0 in $T_P M$ with a neighborhood of P in M. Let $g_0 = g_P$. Fix $v \in T_P M$ so that $g_0(v, v) = \varepsilon = \pm 1$. Let $w \in T_P M$ be such that $\{v, w\}$ forms an orthonormal set relative to g_0. We form the geodesic $\gamma(s) := sv$. Let $e(s)$ be a parallel vector field along γ with $e(0) = w$; we regard e as a vector valued map from a neighborhood of 0 in $\mathbb{R}$ to $T_P M$. Set

$$
\theta(v, s) := \left\{
\begin{array}{ll}
|c|^{-1} \sin(|c|s)e(s) & \text{if } c\varepsilon > 0 \\
se(s) & \text{if } c = 0 \\
|c|^{-1} \sinh(|c|s)e(s) & \text{if } c\varepsilon < 0
\end{array}
\right\}.
$$

Let $\{e_1, \ldots, e_m\}$ be an orthonormal basis for $T_P M$ where $e_1 = v$. We wish to show that:

$$g_{sv}(e_i, e_j) = \begin{cases} 0 & \text{if } i \neq j \\ g_0(v, v) & \text{if } i = j = 1 \\ s^{-2}\theta(v, s)^2 g_0(e_i, e_i) & \text{if } i = j > 1 \end{cases}. \tag{1.10.f}$$

This determines the metric away from the origin and off the *light cone* of null vectors; the metric on the light cone can then be determined by continuity. The lemma will then follow by choosing an isometry between $T_P M$ and $T_{\tilde{p}}\tilde{M}$; this is possible as M and $\tilde{M}$ have the same signature.

We establish Equation (1.10.f) as follows. Examine the Jacobi vector field $Y := sw$ on γ arising from the geodesic spray $T(t, s) := s(v + tw)$. We use the Levi-Civita connection to covariantly differentiate vector fields—thus, for example, $\ddot{\gamma} = 0$ since γ is a geodesic and $\ddot{Y} = -\mathcal{J}(\dot{\gamma})Y$ since Y is a Jacobi vector field. We compute:

$$\partial_s \partial_s g_{\gamma(s)}(\dot{\gamma}(s), Y(s)) = \partial_s g_{\gamma(s)}(\ddot{\gamma}(s), Y(s)) + \partial_s g_{\gamma(s)}(\dot{\gamma}(s), \dot{Y}(s))$$
$$= 0 + \partial_s g_{\gamma(s)}(\dot{\gamma}(s), \dot{Y}(s)) = g_{\gamma(s)}(\ddot{\gamma}(s), \dot{Y}(s)) + g(\dot{\gamma}(s), \ddot{Y}(s))$$
$$= 0 - g_{\gamma(s)}(\dot{\gamma}(s), \mathcal{J}(\dot{\gamma})Y(s)) = -R(Y(s), \dot{\gamma}(s), \dot{\gamma}(s), \dot{\gamma}(s)) = 0.$$

Since $g_{\gamma(s)}(\dot{\gamma}(s), Y(s))(0) = 0$ and $\{\partial_s g_{\gamma(s)}(\dot{\gamma}(s), Y(s))\}(0) = g_0(v, w) = 0$, we may conclude that $Y(s) \perp \dot{\gamma}(s)$ for all s. By Assertion (2) of Lemma 1.47, we have:

$$\ddot{Y} = -\mathcal{J}(\dot{\gamma})Y = -c\varepsilon Y \text{ with } Y(0) = 0 \text{ and } \dot{Y}(0) = w. \tag{1.10.g}$$

Then $Z(s) := \theta(v, s)e(s)$ satisfies the same ordinary differential equation given in Equation (1.10.g) that is satisfied by Y and consequently $Z(s) = Y(s) = sw$. Note that

$$g_{\gamma(s)}(e(s), e(s)) = g(e(0), e(0)) = g_0(w, w).$$

Thus, we have $g_{\gamma(s)}(sw, sw) = \theta(v, s)^2 g_0(w, w)$ if $g_0(v, w) = 0$. Equation (1.10.f) now follows by polarization. ☐

We have actually proved a bit more. By applying the previous lemma to M itself, we have shown that (M, g) is a two-point homogeneous space, i.e., the local isometries of (M, g) act transitively on the pseudo-sphere bundles

$$S^{\pm}(M, g) := \{X \in TM : g(X, X) = \pm 1\}.$$

Let $(V, \langle \cdot, \cdot \rangle)$ be an inner product space of signature (p, q). Let $c > 0$. The *pseudo-spheres*

$$S^{\pm}(V, \langle \cdot, \cdot \rangle, c) := \{\xi \in V : \langle \xi, \xi \rangle = \pm c^2\}$$

have constant sectional curvature $\pm c$. Thus, they provide examples with constant sectional curvature c in all possible signatures.

CONSTANT HOLOMORPHIC SECTIONAL CURVATURE

The fact that a (para)-Kähler manifold $(M, g, J_\pm)$ of dimension $m \geq 4$ and of constant sectional curvature is necessarily flat motivates the investigation of other sectional curvatures more "adapted" to the (para)-Kähler structure. A plane π is said to be a *(para)-holomorphic line* if it is left-invariant by the almost (para)-complex structure, i.e., $J_\pm \pi \subset \pi$. If π is non-degenerate, it has signature $(0, 2)$ or $(2, 0)$ in the complex setting and signature $(1, 1)$ in the para-complex setting. The restriction of the sectional curvature of an almost (para)-Hermitian manifold to the set of (para)-holomorphic lines is called the (para)-holomorphic sectional curvature.

We work first in the algebraic setting. Let $\mathfrak{M} := (V, \langle \cdot, \cdot \rangle, J_\pm, A)$ be a *(para)-Hermitian curvature model*. If $\langle x, x \rangle \neq 0$, let $\pi(x) := \mathrm{Span}\{x, J_\pm x\}$ be the associated non-degenerate *(para)-holomorphic line*. The *(para)-holomorphic sectional curvature* of π is given by:

$$H(\pi) := \mp \langle x, x \rangle^{-2} A(x, J_\pm x, J_\pm x, x).$$

Unlike the sectional curvature, the (para)-holomorphic sectional curvature does not determine the whole curvature tensor in general—we refer to the discussion in Brozos-Vázquez, García-Río, and Gilkey [29]. We are primarily interested in the (para)-Kähler setting—$\mathfrak{M}$ is said to be a *(para)-Kähler curvature model* if the (para)-Kähler identity of Equation (1.4.a) is satisfied:

$$\mathcal{A}(x, y) J_\pm = J_\pm \mathcal{A}(x, y) \text{ or equivalently } A(x, y, J_\pm z, J_\pm w) = \mp A(x, y, z, w).$$

Then the (para)-holomorphic sectional curvature determines the curvature (see Kobayashi and Nomizu [140]). Lemma 1.47 generalizes to this setting to become (see, for example, the discussion in Gilkey [96]):

Lemma 1.50   *Let $\mathfrak{M} := (V, \langle \cdot, \cdot \rangle, J_\pm, A)$ be a (para)-Kähler curvature model. The following conditions are equivalent and, if either is satisfied, then $\mathfrak{M}$ is said to have* constant (para)-holomorphic sectional curvature:

1. *There exist constants λ_0 and λ_1 so*

$$\mathcal{A}(x, y)z = \lambda_0\{\langle y, z \rangle x - \langle x, z \rangle y\} + \lambda_1\{\langle J_\pm y, z \rangle J_\pm x - \langle J_\pm x, z \rangle J_\pm y - 2\langle J_\pm x, y \rangle J_\pm z\}.$$

2. *There exists a constant c so that $H(\pi) = c$ for any (para)-holomorphic line in V.*

If $(V, \langle \cdot, \cdot \rangle, J_\pm, A)$ is a (para)-Kähler curvature model of constant (para)-holomorphic sectional curvature and if x is a spacelike vector, then it follows from Lemma 1.50 that

$$\mathcal{J}(x)y = \left\{ \begin{array}{ll} 0 & \text{if } y = x \\ (\lambda_0 \mp 3\lambda_1)y & \text{if } y = J_\pm x \\ \lambda_0 y & \text{if } y \perp \{x, J_\pm x\} \end{array} \right\}.$$

If the inner product $\langle \cdot, \cdot \rangle$ is positive definite, the sectional curvature and hence, the holomorphic sectional curvature are necessarily bounded, as noted previously in Remark 1.48. However, the failure of the Grassmannian of non-degenerate (para)-holomorphic planes to be compact in general causes the (para)-holomorphic sectional curvature to be unbounded. The (para)-holomorphic sectional curvature is bounded if and only if the model is Bochner flat or, equivalently, if $A(u, J_{\pm}u, J_{\pm}u, u) = 0$ for all null vectors $u \in (V, \langle \cdot, \cdot \rangle)$ (see, for example, Bonome *et al.* [25, 27]). The condition is not that the curvature model has constant (para)-holomorphic sectional curvature.

We say that a (para)-Hermitian manifold is a *(para)-complex space form* if the curvature has the form given in Lemma 1.50 (1), where $\lambda_i = \lambda_i(P)$. Let $(M, g, J_{\pm})$ be such a manifold with $\dim(M) = m$. The case $m = 2$ corresponds to that of a Riemann surface; λ_1 plays no role and there is no constraint on λ_0. The case $m = 4$ is exceptional. If $\lambda_1(P) \equiv 0$, then (M, g) is a space form, so we shall suppose this does not happen. In higher dimensions, if $m \geq 6$, then, in fact, the coefficients λ_i are constant and $\lambda_0 \pm \lambda_1 = 0$. Furthermore, the metric in question is (para)-Kähler.

Note that there are manifolds with pointwise constant holomorphic sectional curvature where the constant varies with the point (see, for example, Gray [117], Gray and Vanhecke [119], and Olszak [171]). We suppose $(M, g, J_{\pm})$ is a (para)-Kähler manifold of constant (para)-holomorphic sectional curvature; the local structure of such manifolds is known (see, for example, Hawley [122] and Igusa [131]). In the Riemannian setting, they are locally modeled on complex projective space or its non-compact dual.

We follow the discussion in Gilkey [96] to construct examples. Let $(V, \langle \cdot, \cdot \rangle, J_-)$ be a pseudo-Hermitian vector space of signature (p, q), where necessarily p and q are even. Let $S^+(V, \langle \cdot, \cdot \rangle)$ be the pseudo-sphere of unit spacelike vectors in V. The unit circle S^1 acts on $S^+(V, \langle \cdot, \cdot \rangle)$ by complex multiplication and we let $\mathbb{C}\mathbb{P}^{p-2,q}$ be the quotient space; this is the set of spacelike complex lines in V. This defines the *Hopf fibration*

$$\pi : S^+(V, \langle \cdot, \cdot \rangle) \rightarrow \mathbb{C}\mathbb{P}^{p-2,q} .$$

The *Fubini–Study metric* g_{FS} on $\mathbb{C}\mathbb{P}^{p-2,q}$ makes π into a *pseudo-Riemannian submersion*; π_* is an isometry from the horizontal subspaces to the tangent space of the base. The construction is essentially the same in the para-complex setting to define the para-complex projective space $\widetilde{CP}^{p,p}$ and the Fubini–Study metric. One has the following classification result—we refer to Barros and Romero [10] and to Kobayashi and Nomizu [140] in the complex setting, and to Cruceanu, Fortuny, and Gadea [60] and to Gadea and Montesinos-Amilibia [85] in the para-complex setting:

Theorem 1.51   *Let (M, g) be a (para)-complex space form of dimension $m \geq 6$. In the complex setting, (M, g) is locally holomorphically isometric to $(\mathbb{C}\mathbb{P}^{u,v}, \lambda g_{FS})$ for some (λ, u, v), while in the para-complex setting, (M, g) is locally para-holomorphically isometric to $(\widetilde{\mathbb{C}\mathbb{P}}^n, \lambda g_{FS})$ for some λ.*

The situation is again special in dimension four. There exist almost (para)-Hermitian manifolds whose curvature tensor has the form of Lemma 1.50 for non-constant functions λ_0 and λ_1. Restrict first to the positive definite almost Hermitian setting and let (M, g, J_-) be such a

4-dimensional almost Hermitian manifold. Any such metric is necessarily Einstein and self-dual. Moreover, the self-dual Weyl curvature operator has exactly two distinct eigenvalues, one with multiplicity two at each point where $\lambda_1 \neq 0$. Then, in a neighborhood of any such point, the metric $\tilde{g} = \{24\|\mathcal{W}^+\|\}^{1/3} g$ is Kähler—see Derdzinski [64]. Hence (M, g, J_-) is locally conformally equivalent to a self-dual Kähler surface (a Bochner flat Kähler surface). This implies, in particular, that the almost complex structure J_- is integrable. Conversely, any Bochner flat Kähler surface with non-constant scalar curvature gives rise to a Hermitian surface with curvature tensor of the form given in Lemma 1.50 by a conformal deformation defined in terms of $\|\mathcal{W}^+\|$ (see Olszak [171]). The almost para-complex case and the almost Hermitian case of higher signature are analogous. They depend on extensions of the generalized Goldberg–Sachs theorem (see Apostolov [8] and Ivanov and Zamkovoy [135]) and the diagonalizability of the self-dual or anti-self-dual Weyl curvature operator (see Cortés-Ayaso, Díaz-Ramos, and García-Río [59]).

1.11   THE SPECTRAL GEOMETRY OF THE CURVATURE TENSOR

In this section, we discuss the spectral geometry of the Jacobi operator, of the skew-symmetric curvature operator, and of the higher order Jacobi operator.

OSSERMAN MANIFOLDS

Osserman [173] initiated the study of the spectral geometry of the Jacobi operator and, for this reason, his name has become attached to the subject. One says that (M, g) is *timelike Osserman* (resp. *spacelike Osserman*) at P if the eigenvalues of the Jacobi operator $\mathcal{J}(X) : Y \to \mathcal{R}(Y, X)X$ are constant on $S^-(T_P M, g_P)$ (resp. $S^+(T_P M, g_P)$). As these are equivalent conditions provided that $p > 0$ and $q > 0$ (see García-Río *et al.* [89]), one simply says that (M, g) is *Osserman* at P. If this condition holds for every point P of M, then (M, g) is said to be *pointwise* Osserman; if the spectrum is independent of the point P, one either adds the words "globally" or omits the modifier "pointwise" entirely.

Note that the Jacobi operator $\mathcal{J}(X)$ is self-adjoint. In the Riemannian setting, Osserman metrics are essentially classified. Let $(V, \langle \cdot, \cdot \rangle)$ be a Riemannian (i.e., positive definite) inner product space of dimension m. Let

$$\mathcal{A}_0(x, y)z := \langle y, z \rangle x - \langle x, z \rangle y$$

be the curvature operator of constant sectional curvature appearing in Lemma 1.47. More generally, if Ψ is a skew-symmetric linear operator on $(V, \langle \cdot, \cdot \rangle)$, then we may define an algebraic curvature tensor by setting:

$$\mathcal{A}_\Psi(x, y)z := \langle \Psi y, z \rangle \Psi x - \langle \Psi x, z \rangle \Psi y - 2\langle \Psi x, y \rangle \Psi z.$$

If $\Psi^2 = -\operatorname{Id}$, then this tensor plays an important role in the analysis of Lemma 1.50. We say that a collection $\{\Psi_1, \ldots, \Psi_k\}$ of skew-symmetric endomorphisms of $(V, \langle \cdot, \cdot \rangle)$ forms a Clif(k) module structure if it satisfies the *Clifford commutation relations*:

$$\Psi_i \Psi_j + \Psi_j \Psi_i = -2\delta_i^j \, \mathrm{Id} \ .$$

The following result of Nikolayevsky [162, 163] uses work of Chi [53] and of Gilkey, Swann, and Vanhecke [114] and summarizes the best results available at present; note that the exclusion of the case of dimension 16 is essential since the Cayley plane is Osserman, but its curvature tensor is not of this type.

Theorem 1.52 *Let* $(V, \langle \cdot, \cdot \rangle, A)$ *be an m-dimensional Riemannian Osserman curvature model with* $m \neq 16$. *Then there exists a* $\mathrm{Clif}(k)$ *module structure* $\{\Psi_1, \ldots, \Psi_k\}$ *on V and there exist constants* α_i *so that* $A = \alpha_0 A_0 + \sum \alpha_i A_{\Psi_i}$.

The possible $\mathrm{Clif}(k)$ module structures are greatly restricted by m; if m is odd, then $(V, \langle \cdot, \cdot \rangle, A)$ does not admit a Clifford module structure and A is necessarily of constant sectional curvature. If $m \equiv 2 \mod 4$, then $k = 1$ and the tensor takes the form given in Lemma 1.50, where $\Psi = J_-$. If $m \equiv 4 \mod 8$, then $k \leq 3$. We refer to work of Adams [2] for further details.

The rank-one symmetric spaces are the space forms, the complex space forms, the quaternionic space forms, the Cayley plane, and their negative curvature duals. All these examples are Osserman. The space forms correspond to $k = 0$, the complex space forms correspond to $k = 1$, and the quaternionic space forms correspond to $k = 3$ in Theorem 1.52; the Cayley plane is exceptional and does not fit into this pattern.

In the Lorentzian setting, Osserman metrics have constant sectional curvature (we refer to Blažić, Bokan, and Gilkey [16] and to García-Río, Kupeli, and Vázquez-Abal [88]). In the higher signature setting, however, the situation is very different and there is a vast literature on the subject— the following are only a few of the possible references and are included to give a flavor of the subject [26, 42, 55, 56, 70, 90, 91], much of it in the 4-dimensional setting [17, 44, 71, 88, 92]. Here, one must emphasize that the spectrum does not determine a self-adjoint operator in the pseudo-Riemannian setting, due to the lack of diagonalizability. Hence, one must pay attention to the possibly non-trivial Jordan normal form of the Jacobi operators and a pseudo-Riemannian manifold (M, g) is said to be *timelike Jordan–Osserman* (resp., *spacelike Jordan–Osserman*) if the Jordan normal form of the Jacobi operators is constant, respectively, on the pseudo-sphere bundles. Although the timelike and the spacelike Osserman conditions are equivalent, the corresponding Jordan–Osserman conditions are not equivalent in the generic situation (see Gilkey [96] for more details). Moreover, the timelike and spacelike Jordan–Osserman conditions imply that the Jacobi operators are diagonalizable if the signature is not neutral, but the Jacobi operators can be rather arbitrary in the neutral signature case (see the work of Gilkey and Ivanova [99, 100] for further details).

The Osserman condition is extended in Section 2.3 to the affine setting. It turns out that an m-dimensional affine manifold (M, ∇) is *affine Osserman* at point P of M if $\mathcal{J}(X)$ is nilpotent for all $X \in T_P M$, or, equivalently, if $\{0\}$ is the only eigenvalue of $\mathcal{J}(X)$ or, equivalently,

$$\mathcal{J}(X)^m = 0 \text{ for all } X \in T_P M \ .$$

If this condition holds at every point P of M, then (M, ∇) is said to be *affine Osserman*. We refer, for example, to the discussion in García-Río *et al.* [89] for further details. Note that this definition is perfectly natural in the context of Weyl geometry.

One can also study similar conditions that are determined by the conformal curvature operator rather than the curvature operator itself. In this setting, a pseudo-Riemannian manifold (M, g) is said to be *conformally Osserman* if for each point $P \in M$ the spectrum of the Jacobi operator $\mathcal{J}_W(X)$ corresponding to the Weyl conformal curvature tensor is constant on the corresponding unit pseudo-spheres $S^{\pm}(T_P M, g_P)$. This condition is meaningless if M is 3-dimensional (since the Weyl conformal curvature tensor vanishes). Furthermore, in dimension four, this condition is equivalent to the manifold being (anti)-self-dual (see, for example, Blažić *et al.* [19] and Brozos-Vázquez, García-Río, and Vázquez-Lorenzo [33]). Riemannian conformally Osserman manifolds have been recently studied by Nikolayevsky [164, 165]. Note that one of the main difficulties in working with the conformal Osserman property arises from the fact that the Weyl conformal curvature tensor does not satisfy the second Bianchi identity.

THE HIGHER ORDER JACOBI OPERATOR

Following the seminal work of Stanilov and Videv [197], one can define a *higher order Jacobi operator*. Let $\{e_1, \ldots, e_k\}$ be a basis for a non-degenerate k-plane π of signature (u, v). Set:

$$\mathcal{J}(\pi) : X \to \sum_{i,j} g^{ij} \mathcal{R}(X, e_i) e_j .$$

This operator is independent of the particular basis chosen. If $\{e_i\}$ are an orthonormal basis, then

$$\mathcal{J}(\pi) := \varepsilon_1 \mathcal{J}(e_1) + \cdots + \varepsilon_k \mathcal{J}(e_k) \text{ where } \varepsilon_i := g(e_i, e_i) . \tag{1.11.a}$$

We refer to Gilkey [96] and to Gilkey, Nikčević, and Videv [110] for further details concerning this operator. If $k = 1$, this is the normalized Jacobi operator while, if $k = m$, this is the Ricci operator. Let $\mathrm{Gr}_{u,v}$ denote the Grassmannian of non-degenerate linear subspaces of TM of signature (u, v). We suppose $0 \le u \le p$ and $0 \le v \le q$. We say that (M, g) is *higher order Osserman of type* (u, v) if the spectrum of $\mathcal{J}(\pi)$ is constant on $\mathrm{Gr}_{u,v}$; if (u, v) and $(\tilde{u}, \tilde{v})$ are admissible and if $u + v = \tilde{u} + \tilde{v}$, then (M, g) is higher order Osserman of type (u, v) if and only if (M, g) is higher order Osserman of type $(\tilde{u}, \tilde{v})$ and, hence, (M, g) is simply said to be *k-higher order Osserman* where $k = u + v$. This condition is very restrictive in the Riemannian and Lorentzian settings since any k-higher order Osserman manifold is necessarily of constant sectional curvature for $2 \le k \le m - 2$; we refer to Gilkey [97] in the Riemannian setting and to Gilkey and Stavrov [113] in the Lorentzian setting for further details. This operator will play an important role in Section 3.5.

THE SKEW-SYMMETRIC CURVATURE OPERATOR

Let (M, g) be a pseudo-Riemannian manifold and let ∇ be an affine connection on M. We shall usually take $\nabla = {}^g\nabla$ to be the Levi-Civita connection, but, again, this is not strictly necessary and the questions involved are natural in the context of Weyl geometry. Let π be a non-degenerate oriented 2-plane in $T_P M$. If $\{x, y\}$ is an oriented basis for π, the *skew-symmetric curvature operator* is defined by setting

$$\langle \mathcal{R}(\pi)z, w \rangle = |g(x, x)g(y, y) - g(x, y)^2|^{-1/2} R(x, y, z, w). \qquad (1.11.b)$$

It is independent of the particular oriented basis chosen. Following the seminal work of Ivanov and Petrova [134], one says (M, g, ∇) is *spacelike Ivanov–Petrova* (resp. *timelike Ivanov–Petrova* or *mixed Ivanov–Petrova*) at P if the eigenvalues of $\mathcal{R}(\cdot)$ are constant on the Grassmannian of oriented spacelike (resp. timelike or mixed) 2-planes in $T_P M$. Again, these are equivalent conditions so one simply speaks of (M, g, ∇) being *Ivanov–Petrova* at P; if this condition holds for all points P, (M, g, ∇) is simply said to be *Ivanov–Petrova* or *globally Ivanov–Petrova*; if the eigenvalues vary with the point in question, the adjective *pointwise* is often used. In the higher signature setting, there can be non-trivial Jordan normal form and if the Jordan normal form is constant, one adds the adjective *Jordan*. There is a vast literature on the subject and again, we can only cite a few references [41, 44, 110, 115, 216].

CHAPTER 2

The Geometry of Deformed Riemannian Extensions

Let M be a smooth manifold. In this chapter, we shall consider some of the geometrical aspects of a Riemannian extension g_∇ or of a deformed Riemannian extension $g_{\nabla,\Phi}$, which provide a link between the affine geometry of (M, ∇) and the neutral signature pseudo-Riemannian geometry of T^*M. We shall investigate the spectral geometry of the Jacobi operator and of the skew-symmetric curvature operator both on M and on T^*M. Here is a brief outline to Chapter 2. Section 2.1 provides a brief summary of some of the relevant material we shall need in the chapter. Section 2.2 gives a construction that yields a wide variety of affine Osserman and affine Ivanov–Petrova manifolds. Section 2.3 deals with affine surfaces; the geometric properties of affine Osserman surfaces and of affine Ivanov–Petrova surfaces are examined in some detail. Section 2.4 examines homogeneous affine connections in the context of 2-dimensional geometry. Finally, Section 2.5 deals with the higher-dimensional setting.

2.1 BASIC NOTATIONAL CONVENTIONS

In this section, we review the notational conventions and basic constructions which underlie Chapter 2. We refer to Chapter 1 for further details.

SPECIAL METRICS

Let (M, ∇) be an *affine manifold* of dimension $\bar{m}$; this means that ∇ is a torsion-free connection on M. In local coordinates $\vec{x} = (x^1, \ldots, x^{\bar{m}})$, the Christoffel symbols are defined by the identity:

$$\nabla_{\partial_{x_i}} \partial_{x_j} = \Gamma_{ij}{}^k \partial_{x_k} \text{ where } \Gamma_{ij}{}^k = \Gamma_{ji}{}^k.$$

We adopt the notation of Equation (1.1.d) and introduce dual fiber coordinates $x_{i'}$ on the cotangent bundle by decomposing a 1-form as $\omega = x_{i'} dx^i$. The *Riemannian extension* of Definition 1.32 is the neutral signature metric on T^*M, which is given in local coordinates relative to the frame $\{\partial_{x_1}, \ldots, \partial_{x_{\bar{m}}}, \partial_{x^{1'}}, \ldots, \partial_{x^{\bar{m}'}}\}$ by:

$$g_\nabla = \begin{pmatrix} -2x_{k'}\Gamma_{ij}{}^k(\vec{x}) & \text{Id}_{\bar{m}} \\ \text{Id}_{\bar{m}} & 0 \end{pmatrix}.$$

More generally, if Φ is a symmetric $(0, 2)$-tensor field on M, then the *deformed Riemannian extension* $g_{\nabla,\Phi}$ is the metric of neutral signature on T^*M given by Equation (1.7.a):

$$g_{\nabla, \Phi} = \begin{pmatrix} -2x_{k'}\Gamma_{ij}{}^k(\vec{x}) + \Phi_{ij}(\vec{x}) & \mathrm{Id}_{\bar{m}} \\ \mathrm{Id}_{\bar{m}} & 0 \end{pmatrix}.$$

These metrics are invariantly defined and are independent of the particular coordinate system $(x^1, \ldots, x^{\bar{m}})$ chosen on the base. We say that a neutral signature pseudo-Riemannian manifold (N, g_N) is a *Walker metric* if, relative to some system of local coordinates, we have

$$g_N = \begin{pmatrix} B & \mathrm{Id}_{\bar{m}} \\ \mathrm{Id}_{\bar{m}} & 0 \end{pmatrix} \text{ for } 2\bar{m} = \dim(N).$$

Thus, in particular, if B is a polynomial of order at most 1 in the $x_{i'}$ variables, then g is locally a deformed Riemannian extension; a deformed Riemannian extension is locally a Riemannian extension if B vanishes on the "zero-section". In these two instances, the linear terms in the $x_{i'}$ variables give the connection 1-form of a torsion-free connection on the base manifold.

NATURAL OPERATORS DEFINED BY THE CURVATURE

Let $\mathcal{R}$ be the curvature operator corresponding to the Levi-Civita connection of a pseudo-Riemannian manifold (M, g) and let R be the associated curvature tensor as defined in Equation (1.2.b) and in Equation (1.2.d), respectively:

$$\mathcal{R}(X, Y)Z := (\nabla_X \nabla_Y - \nabla_Y \nabla_X - \nabla_{[X,Y]})Z \text{ and } R(X, Y, Z, W) := g(\mathcal{R}(X, Y)Z, W).$$

The Jacobi operator of Equation (1.10.a) is described by:

$$\mathcal{J}(X) : Y \to \mathcal{R}(Y, X)X \text{ and } g(\mathcal{J}(X)Y, Z) = R(Y, X, X, Z).$$

We use the symmetries of the curvature tensor given in Equations (1.2.l)-(1.2.o) to see that $\mathcal{J}(X)$ is self-adjoint since $R(Y, X, X, Z) = R(Z, X, X, Y)$. In the positive definite setting, this implies that $\mathcal{J}(X)$ is diagonalizable, but this is not the case in the indefinite signature setting and is the reason why these two contexts are so different. Similarly, let $\{x, y\}$ be an oriented basis for a non-degenerate 2-plane $\pi := \mathrm{Span}\{x, y\}$. The *skew-symmetric curvature operator* of Equation (1.11.b) is then given by:

$$\mathcal{R}(\pi) := |g(x, x)g(y, y) - g(x, y)^2|^{-1/2}\mathcal{R}(x, y) \text{ so}$$
$$g(\mathcal{R}(\pi)z, w) = |g(x, x)g(y, y) - g(x, y)^2|^{-1/2}R(x, y, z, w).$$

OSSERMAN GEOMETRY

If J is a linear operator, then the *spectrum* $\mathrm{Spec}\{J\}$ is the set of (possibly complex) eigenvalues of J. Let $p_\lambda(J) := \det\{\lambda\,\mathrm{Id} - J\}$ be the *characteristic polynomial* of J; λ belongs to $\mathrm{Spec}\{J\}$ if and only if $p_\lambda(J) = 0$. One has the following useful observation; we shall omit the proof as it is an elementary exercise in linear algebra using Jordan normal form.

Lemma 2.1   *Let J be a linear map of an m-dimensional vector space V. The following assertions are equivalent and, if any is satisfied, then J is said to be nilpotent.*

1. $\mathrm{Spec}\{J\} = \{0\}$.

2. $J^m = 0$.

3. $p_\lambda(J) = \lambda^m$.

4. $\mathrm{Tr}\{J^k\} = 0$ for $1 \le k \le m$.

Recall that a pseudo-Riemannian manifold (M, g) is said to be pseudo-Riemannian *Osserman* (if we wish to emphasize the fact that we are in the pseudo-Riemannian setting) or simply *Osserman* (if the context is clear), provided that $\mathrm{Spec}\{\mathcal{J}(\cdot)\}$ is constant on the pseudo-sphere bundle $S^+(M, g)$ of unit spacelike tangent vectors if $q > 0$ or, equivalently, on the pseudo-sphere bundle $S^-(M, g)$ of unit timelike tangent vectors if $p > 0$. This implies that the eigenvalue multiplicities are constant as well and, consequently, (M, g) is Osserman if and only if the characteristic polynomial of $\mathcal{J}(X)$ is constant on $S^+(M, g)$ or on $S^-(M, g)$ (the polynomials can be different). In the positive definite setting, Osserman [173] wondered if such a manifold must either be flat or be locally isometric to a rank-one symmetric space. This result was proved by Chi [53] if $m \not\equiv 0 \mod 4$—see Gilkey, Swann, and Vanhecke [114] for related work and additional references. Any Osserman Lorentzian manifold has constant sectional curvature (see Blažić, Bokan, and Gilkey [16] and also García-Río, Kupeli, and Vázquez-Abal [88]). However, the situation is completely different for metrics of indefinite signature. If $p \ge 2$ and if $q \ge 2$, then there are Osserman pseudo-Riemannian manifolds of signature (p, q) that are not locally symmetric (see, for example, García-Río, Vázquez-Abal, and Vázquez-Lorenzo [91]).

Let (M, ∇) be an affine manifold of dimension $\bar{m}$. We say that (M, ∇) is *affine Osserman* if $\mathrm{Spec}\{\mathcal{J}(X)\} = \{0\}$ for all X, or, equivalently, if $\mathcal{J}(X)^{\bar{m}} = 0$ for any tangent vector X. Let Φ be an arbitrary symmetric $(0, 2)$-tensor field on M. We will show in Theorem 2.15 that (M, ∇) is affine Osserman if and only if the deformed Riemannian extension $(T^*M, g_{\nabla, \Phi})$ is pseudo-Riemannian Osserman.

IVANOV–PETROVA GEOMETRY

A pseudo-Riemannian manifold (M, g) is said to be *Ivanov–Petrova* at a point P of M if $\mathrm{Spec}\{\mathcal{R}(\pi)\}$ is constant on the Grassmannian $\mathrm{Gr}^+_{u,v}(T_P M)$ of oriented 2-planes of signature (u, v); this notion is independent of whether (u, v) is taken to be $(2, 0)$ (if $p \ge 2$), or $(1, 1)$ (if $p \ge 1$ and $q \ge 1$), or $(0, 2)$ (if $q \ge 2$). Metrics of constant curvature are Ivanov–Petrova, but there are Ivanov–Petrova metrics that do not have constant sectional curvature (we refer to Gilkey [96], to Gilkey, Leahy, and Sadofsky [101] and to Gilkey and Zhang [115]). For example, let $M = I \times N$ be a product manifold where I is a subinterval of $\mathbb{R}$ and where ds_N^2 is a metric of constant sectional curvature K on a pseudo-Riemannian manifold N of dimension $m - 1$. Give M the metric

$$ds_M^2 := dt^2 + f(t)ds_N^2 \quad \text{where} \quad f(t) := \tfrac{1}{2}(Kt^2 + At + B) \ne 0.$$

Then M is a pseudo-Riemannian Ivanov–Petrova manifold; if $A^2 - 4BK \neq 0$, then (M, g) does not have constant sectional curvature. Conversely, if $m \geq 9$, then any Riemannian Ivanov–Petrova metric either has constant sectional curvature or is locally isometric to this example, where $f(t) > 0$ and N is Riemannian.

Ivanov–Petrova metrics which are 3-dimensional have been investigated in the Riemannian and Lorentzian setting (see García-Río, Haji-Badali, and Vázquez-Lorenzo [87] and Ivanov and Petrova [134]), where a complete algebraic description is available. A 3-dimensional pseudo-Riemannian manifold is Ivanov–Petrova if and only if at each point P in M where the sectional curvature is non-constant, either the Ricci operator is diagonalizable of rank one, or the Ricci operator is 2-step nilpotent (see [41, 87, 134, 161]). We say an affine manifold (M, ∇) of dimension $\bar{m}$ is *affine Ivanov–Petrova* at a point P if

$$\text{Spec}\{\mathcal{R}(x, y)\} = \{0\} \text{ or equivalently if } \mathcal{R}(x, y)^{\bar{m}} = 0 \text{ for all } x, y \in T_P M.$$

Let Φ be an arbitrary symmetric $(0, 2)$-tensor field on M. We will show in Theorem 2.15 that (M, ∇) is affine Ivanov–Petrova if and only if the deformed Riemannian extension $g_{\nabla, \Phi}$ on the cotangent bundle is pseudo-Riemannian Ivanov–Petrova. We shall characterize those affine connections on surfaces that are affine Ivanov–Petrova in terms of the symmetry and degeneracy of their Ricci tensor. A local description of recurrent Ivanov–Petrova connections on surfaces will be given in Theorem 2.7.

HOMOGENEOUS AFFINE CONNECTIONS

The study of homogeneous affine connections on surfaces which are either Osserman or Ivanov–Petrova is more involved and relies on a classification result by Opozda [172] (see also Kowalski, Opozda, and Vlášek [143, 144]). Any homogeneous affine connection is either the Levi-Civita connection of a metric of constant curvature or it belongs to one of two families A and B (see Theorem 2.8). We shall consider the equivalence problem for such connections, showing that, in addition to the flat ones, there is an explicit family of locally homogeneous affine connections that is of both Types A and B. This family is contained in the class of projectively flat and recurrent homogeneous connections with symmetric and degenerate Ricci tensor (see Theorem 2.13).

PROJECTIVELY AFFINE OSSERMAN MANIFOLDS AND PROJECTIVELY AFFINE IVANOV–PETROVA MANIFOLDS

We have $\mathcal{J}(cx) = c^2 \mathcal{J}(x)$ and $\mathcal{R}(cx, cy) = c^2 \mathcal{R}(x, y)$. In the pseudo-Riemannian setting, we can use the metric to normalize x to belong to $S^{\pm}(M, g)$ or to normalize $\{x, y\}$ to be an orthonormal set and eliminate the effect of this rescaling. This renormalization is not available in the affine setting and thus, it is natural to consider the setting in that $\text{Spec}\{\mathcal{J}(x)\} = \{0\}$ or $\text{Spec}\{\mathcal{R}(x, y)\} = \{0\}$. There is, however, another notion available in the affine setting that has not been studied extensively in the literature. We say (M, ∇) is *projectively affine Osserman* at a point $P \in M$ if, given any pair of non-zero vectors $x, y \in T_P M$, there is a non-zero complex number $\lambda(x, y) \neq 0$ so:

$$\text{Spec}\{\mathcal{J}(y)\} = \lambda(x, y) \cdot \text{Spec}\{\mathcal{J}(x)\} \neq \{0\}\,.$$

We explicitly rule out the case where the spectrum solely consists of $\{0\}$, as that is the affine Osserman setting discussed above. Similarly, we say that (M, ∇) is *projectively affine Ivanov–Petrova* if, given a quadruple of vectors $x, y, \bar{x}, \bar{y} \in T_P M$ with $\{x, y\}$ linearly independent and $\{\bar{x}, \bar{y}\}$ linearly independent, then there exists $\lambda(x, y, \bar{x}, \bar{y}) \neq 0$ so:

$$\text{Spec}\{\mathcal{R}(\bar{x}, \bar{y})\} = \lambda(x, y, \bar{x}, \bar{y}) \cdot \text{Spec}\{\mathcal{R}(x, y)\} \neq \{0\}\,.$$

If g is a positive definite Osserman metric that is not flat, then ${}^g\nabla$ is projectively affine Osserman, but is not affine Osserman. Similarly, if g is a positive definite Ivanov–Petrova metric that is not flat, then ${}^g\nabla$ is projectively affine Ivanov–Petrova, but is not affine Ivanov–Petrova. Indefinite Osserman metrics are not projectively affine Osserman since $\text{Spec}\{\mathcal{J}(x)\} = 0$ if x is a null vector. Similarly, indefinite Ivanov–Petrova metrics are not projectively affine Ivanov–Petrova as $\text{Spec}\{\mathcal{R}(x, y)\} = \{0\}$ if the induced metric on $\text{Span}\{x, y\}$ is degenerate. We refer to Gilkey and Nikčević [106] for further details.

Example 2.2   Let $(x^1, \ldots, x^{\bar{m}})$ be coordinates on $\mathbb{R}^{\bar{m}}$ for $\bar{m} \geq 2$. Let $\mathcal{M} = (\mathbb{R}^{\bar{m}}, \nabla)$ be the affine manifold defined by setting:

$$\Gamma_{11}{}^1 = 2, \quad \Gamma_{i1}{}^1 = \Gamma_{1i}{}^i = 1, \text{ and } \Gamma_{ii}{}^1 = 1 \text{ for } i > 1\,.$$

The non-zero components of the curvature tensor are given by

$$R_{ijj}{}^i = 1 \text{ and } R_{jij}{}^i = -1 \text{ for } i \neq j\,.$$

This is the curvature tensor of a metric of constant sectional curvature 1 and hence, is projectively Osserman and projectively Ivanov–Petrova. Since the Christoffel symbols are constant, the group of translations acts transitively on $\mathcal{M}$ by affine isomorphisms; thus, $\mathcal{M}$ is affine homogeneous. However, if we set $\gamma(t) = (x(t), 0, \ldots, 0)$, then the geodesic equation becomes $\ddot{x} = -2\dot{x}x$, which blows up in finite time with initial conditions $x(0) = -1$ and $\dot{x}(0) = -1$. Thus, $\mathcal{M}$ is geodesically incomplete. Finally, since $\nabla \mathcal{R} \neq 0$, these manifolds are not locally symmetric. This shows that the affine manifold $\mathcal{M}$ is not affinely equivalent to a manifold of constant sectional curvature.

2.2   EXAMPLES OF AFFINE OSSERMAN IVANOV–PETROVA MANIFOLDS

Since much of our discussion will be concerned with affine Osserman and affine Ivanov–Petrova manifolds, it is worth giving a construction that exhibits a wide family of such examples in a uniform context (we refer to Gilkey [98] for more details).

Definition 2.3   Let $(x^1, \ldots, x^{\bar{m}})$ be the usual system of coordinates on $M = \mathbb{R}^{\bar{m}}$. We say that an affine manifold (M, ∇) is a *generalized affine plane wave manifold* if

$$\nabla_{\partial_{x_i}} \partial_{x_j} = \sum_{k > \max(i,j)} \Gamma_{ij}{}^k(x^1, \ldots, x^{k-1}) \partial_{x_k} .$$

Here, we adopt the convention that the empty sum is zero. For example, if $\bar{m} = 2$, then the possibly non-zero Christoffel symbols could be $\Gamma_{11}{}^2(x^1)$, while, if $\bar{m} = 3$, the possibly non-zero Christoffel symbols could be:

$$\{\Gamma_{11}{}^2(x^1),\ \Gamma_{11}{}^3(x^1, x^2),\ \Gamma_{12}{}^3(x^1, x^2) = \Gamma_{21}{}^3(x^1, x^2),\ \Gamma_{22}{}^3(x^1, x^2)\}.$$

Theorem 2.4   *Let (M, ∇) be a generalized affine plane wave manifold.*

1. *(M, ∇) is geodesically complete.*

2. *There is a unique geodesic joining any two points of M.*

3. *(M, ∇) is Ricci flat, affine Osserman, and affine Ivanov–Petrova.*

Proof. Let (M, ∇) be a generalized affine plane wave manifold. The geodesic equation takes the form:

$$\ddot{\gamma}^k(t) + \sum_{i,j < k} \dot{\gamma}^i(t) \dot{\gamma}^j(t) \Gamma_{ij}{}^k(\gamma^1, \ldots, \gamma^{k-1})(t) = 0. \qquad (2.2.\text{a})$$

Let $P = (P^1, \ldots, P^{\bar{m}})$ be the initial position and let $\xi = (\xi^1, \ldots, \xi^{\bar{m}})$ be the initial velocity. To solve Equation (2.2.a) with $\gamma(0) = P$ and $\dot{\gamma}(0) = \xi$, we define:

$$\gamma^1(t) := P^1 + \xi^1 t,$$
$$\gamma^k(t) := P^k + \xi^k t - \int_0^t \int_0^s \sum_{i,j<k} \dot{\gamma}^i(r) \dot{\gamma}^j(r) \Gamma_{ij}{}^k(\gamma^1, \ldots, \gamma^{k-1})(r) \, dr \, ds \text{ for } k > 1.$$

This establishes Assertion (1). Given $P, Q \in M$, there is a unique geodesic $\gamma = \gamma_{P,Q}$ with $\gamma(0) = P$ and $\gamma(1) = Q$ where

$$\xi^1 = Q^1 - P^1,$$
$$\xi^k = Q^k - P^k + \int_0^1 \int_0^s \sum_{i,j<k} \dot{\gamma}^i(r) \dot{\gamma}^j(r) \Gamma_{ij}{}^k(\gamma^1, \ldots, \gamma^{k-1})(r) \, dr \, ds \text{ for } k > 1.$$

This establishes Assertion (2). We have by Equation (1.2.c) that:

$$\begin{aligned}
R_{ijk}{}^\ell = {} & \partial_{x_i} \Gamma_{jk}{}^\ell(x^1, \ldots, x^{\ell-1}) - \partial_{x_j} \Gamma_{ik}{}^\ell(x^1, \ldots, x^{\ell-1}) \\
& + \Gamma_{in}{}^\ell(x^1, \ldots, x^{\ell-1}) \Gamma_{jk}{}^n(x^1, \ldots, x^{n-1}) \\
& - \Gamma_{jn}{}^\ell(x^1, \ldots, x^{\ell-1}) \Gamma_{ik}{}^n(x^1, \ldots, x^{n-1}).
\end{aligned}$$

Suppose $\ell \leq k$. Then $\Gamma_{jk}{}^{\ell} = \Gamma_{ik}{}^{\ell} = 0$ so $\partial_{x_i}\Gamma_{jk}{}^{\ell} = \partial_{x_j}\Gamma_{ik}{}^{\ell} = 0$. Furthermore, for either of the quadratic terms to be non-zero, there must exist an index n with $k < n$ and $n < \ell$, which is not possible. Consequently $R_{ijk}{}^{\ell} = 0$ if $\ell \leq k$. Thus for arbitrary $\{\xi_1, \xi_2\}$ we have:

$$\mathcal{R}(\xi_1, \xi_2)\partial_{x_k} \in \mathrm{Span}_{k < \ell}\{\partial_{x_\ell}\}.$$

This shows that $\mathcal{R}$ is nilpotent so (M, ∇) is affine Ivanov–Petrova. Suppose that $\ell \leq i$. Then

$$\partial_{x_i}\Gamma_{jk}{}^{\ell}(x^1, \ldots, x^{\ell-1}) = 0 \quad \text{and} \quad \partial_{x_j}\Gamma_{ik}{}^{\ell} = \partial_{x_j}0 = 0.$$

We have $\Gamma_{in}{}^{\ell} = 0$ for any n. For the other quadratic term to be non-zero, there must exist an index n so $i < n$ and $n < \ell$, which is not possible. This shows $R_{ijk}{}^{\ell} = 0$ if $\ell \leq i$; similarly $R_{ijk}{}^{\ell} = 0$ if $\ell \leq j$. As a result, (M, ∇) is Ricci flat. Furthermore, for any ξ, we have:

$$\mathcal{J}(\xi) : \partial_{x_i} \to \mathrm{Span}_{i < \ell}\{\partial_{x_\ell}\}.$$

This shows that $\mathcal{J}(\xi)$ is nilpotent as well so (M, ∇) is affine Osserman. $\qquad \square$

The following special case is instructive:

Theorem 2.5   *Let (x^1, x^2, x^3) be the usual system of coordinates on $\mathbb{R}^3$ and consider the affine connection ∇ on $\mathbb{R}^3$, whose only non-trivial Christoffel symbol is $\Gamma_{12}{}^3 = \Gamma_{21}{}^3 = x^1$. Then $(\mathbb{R}^3, \nabla)$ is Ricci flat, locally symmetric, affine Osserman, affine Ivanov–Petrova, and not flat.*

Proof.   We apply Theorem 2.4. Since $(\mathbb{R}^3, \nabla)$ is a generalized affine plane wave manifold, it is Ricci flat, affine Osserman, and affine Ivanov–Petrova. There are no non-zero quadratic terms in the expression

$$R_{ija}{}^b = \partial_{x_i}\Gamma_{ja}{}^b - \partial_{x_j}\Gamma_{ia}{}^b + \Gamma_{ic}{}^b\Gamma_{ja}{}^c - \Gamma_{jc}{}^b\Gamma_{ia}{}^c,$$

which is given in Equation (1.2.c), and the only non-zero component of the curvature is

$$R_{121}{}^3 = -R_{211}{}^3 = \partial_{x_1}\Gamma_{21}{}^3 = 1.$$

Consequently $(\mathbb{R}^3, \nabla)$ is not flat. Furthermore, when considering the expression

$$\nabla\mathcal{R}(X_1, X_2; X_4)X_3 = \nabla_{X_4}\mathcal{R}(X_1, X_2)X_3 - \mathcal{R}(\nabla_{X_4}X_1, X_2)X_3$$
$$-\mathcal{R}(X_1, \nabla_{X_4}X_2)X_3 - \mathcal{R}(X_1, X_2)\nabla_{X_4}X_3,$$

which is given in Equation (1.2.g), the terms in $\nabla_{X_4}X_j$ vanish, since $\nabla_{X_4}X_j$ is a multiple of ∂_{x_3} and $\mathcal{R}(Y_1, Y_2)Y_3 = 0$ if any of the $Y_i \in \mathrm{Span}\{\partial_{x_3}\}$. Similarly since $\nabla\partial_{x_3} = 0$, $\nabla_{X_4}\mathcal{R}(X_1, X_2)X_3 = 0$ as well. Thus $\nabla\mathcal{R} = 0$ as desired. $\qquad \square$

2.3 THE SPECTRAL GEOMETRY OF THE CURVATURE TENSOR OF AFFINE SURFACES

In this section, we examine the geometry of affine surfaces (Σ, ∇). We first study the Jacobi operator and then we study the skew-symmetric curvature operator. If $\bar{m} = 2$, then

$$\rho_{11} = R_{111}{}^1 + R_{211}{}^2 = -R_{121}{}^2, \quad \rho_{12} = R_{112}{}^1 + R_{212}{}^2 = -R_{122}{}^2,$$
$$\rho_{21} = R_{121}{}^1 + R_{221}{}^2 = R_{121}{}^1, \quad \rho_{22} = R_{122}{}^1 + R_{222}{}^2 = R_{122}{}^1, \tag{2.3.a}$$

$$\mathcal{R}(e_1, e_2) = \begin{pmatrix} \rho_{21} & \rho_{22} \\ -\rho_{11} & -\rho_{12} \end{pmatrix}, \quad \det\{\lambda\,\mathrm{Id} - \mathcal{R}\} = \lambda^2 + \lambda(\rho_{21} - \rho_{12}) + \det\{\rho\}.$$

THE JACOBI OPERATOR

Let ρ_a and ρ_s be the alternating and symmetric Ricci tensors of an affine connection ∇. Affine connections with skew-symmetric Ricci tensor have been studied extensively in the literature [6, 15, 66, 142, 143, 213].

Theorem 2.6 *Let P be a point of a connected 2-dimensional affine surface (Σ, ∇).*

1. *(Σ, ∇) is affine Osserman at P if and only if $\rho_s = 0$.*

2. *(Σ, ∇) is projectively affine Osserman at P if and only ρ_s is definite.*

3. *If (Σ, ∇) is locally symmetric and affine Osserman, then ∇ is flat.*

4. *Let (Σ, ∇) be affine Osserman. Suppose that $\rho \neq 0$ at $P \in \Sigma$. Then there exist local coordinates (x^1, x^2) defined near P and there exists a smooth function θ that is defined near P with $\partial_{x_1}\partial_{x_2}\theta(P) \neq 0$ so that the (possibly) non-zero covariant derivatives are $\nabla_{\partial_{x_1}}\partial_{x_1} = -(\partial_{x_1}\theta)\partial_{x_1}$ and $\nabla_{\partial_{x_2}}\partial_{x_2} = (\partial_{x_2}\theta)\partial_{x_2}$.*

5. *Let (Σ, ∇) be a compact affine surface with skew-symmetric Ricci tensor. Then ρ is exact and Σ is diffeomorphic to the Klein bottle or to the torus.*

Proof. By definition $\rho(x, x) = \mathrm{Tr}\{\mathcal{J}(x)\}$. Because $\mathcal{J}(x)x = \mathcal{R}(x, x)x = 0$ and because the dimension is two, one of the following two possibilities pertains:

1. $\mathrm{Spec}\{\mathcal{J}(x)\} = \{0\}$ and $\rho(x, x) = 0$.

2. $\mathrm{Spec}\{\mathcal{J}(x)\} = \{0, \lambda(x)\}$ for $0 \neq \lambda(x) \in \mathbb{R}$ and $\rho(x, x) = \lambda(x) \neq 0$.

Assertion (1) now follows as $\rho(x, x) = 0$ for all x if and only if ρ is skew-symmetric. We have ρ is definite if and only if $\rho(x, x) \neq 0$ for $x \neq 0$ or, equivalently, $\mathrm{Spec}\{\mathcal{J}(x)\} \neq \{0\}$ for $x \neq 0$. Assertion (2) now follows.

Let (Σ, ∇) be a connected affine surface with $\nabla \mathcal{R} = 0$. Because the dimension is two, either $\mathrm{Rank}\{\rho_a\} = 0$ or $\mathrm{Rank}\{\rho_a\} = 2$. Since $\nabla \mathcal{R} = 0$, we have $\nabla \rho_a = 0$ and thus, $\mathrm{Rank}\{\rho_a\}$ is constant. Suppose that $\mathrm{Rank}\{\rho_a\} = 2$ at each point. Then ρ_a defines a parallel volume form. By Theorem 1.7, ρ is symmetric, which contradicts the assumption $\mathrm{Rank}\{\rho_a\} = 2$. Consequently, $\mathrm{Rank}\{\rho_a\} = 0$ so ρ is symmetric. But, since (Σ, ∇) is affine Osserman, ρ is skew-symmetric and $\rho = 0$. Assertion (3) now follows from Equation (2.3.a).

We now prove Assertion (4). Let (Σ, ∇) be an affine Osserman surface. Let $\rho \neq 0$ near P. Then ρ is skew-symmetric and defines a volume form in a neighborhood U of P. Therefore ρ is recurrent. The work of Blažić and Bokan [15] shows that ∇ has recurrent curvature tensor. The work of Wong [213] then implies that there is a system of coordinates (x^1, x^2) defined near P so that one of the following three possibilities holds:

1. There is a smooth function θ with $\partial_{x_1} \partial_{x_2} \theta \neq 0$ so that
$$\nabla_{\partial_{x_1}} \partial_{x_1} = -(\partial_{x_1}\theta)\partial_{x_1} \text{ and } \nabla_{\partial_{x_2}} \partial_{x_2} = (\partial_{x_2}\theta)\partial_{x_2}.$$

2. There is a smooth function φ with $\partial_{x_1} \partial_{x_2} \ln \varphi \neq 0$ so that
$$\nabla_{\partial_{x_1}} \partial_{x_1} = -(\partial_{x_1} \ln \varphi)\partial_{x_1} \text{ and } \nabla_{\partial_{x_2}} \partial_{x_2} = \varphi \partial_{x_1} + (\partial_{x_2} \ln \varphi)\partial_{x_2}.$$

3. There is a smooth function ψ with $\partial_{x_1} \partial_{x_2} \ln \psi \neq 0$ so that
$$\nabla_{\partial_{x_1}} \partial_{x_1} = \left(-\partial_{x_1} \ln \psi + \frac{x^2}{1+x^1 x^2}\right)\partial_{x_1} + \left(\frac{1}{\psi(1+x^1 x^2)}\right)\partial_{x_2},$$
$$\nabla_{\partial_{x_2}} \partial_{x_2} = -\left(\frac{\psi}{1+x^1 x^2}\right)\partial_{x_1} + \left(\partial_{x_2} \ln \psi + \frac{x^1}{1+x^1 x^2}\right)\partial_{x_2}.$$

A simplification of Wong's result [213] was obtained by Derdzinski [66], who showed that one can eliminate the final two cases. This completes the proof of Assertion (4); we refer to Derdzinski [66] for the proof of the final assertion. □

We note that Theorem 2.5 shows that Theorem 2.6 (3) fails if $\bar{m} > 2$. We also note that there exist examples of non-flat locally symmetric Osserman pseudo-Riemannian manifolds on $\mathbb{R}^4$ with metric of signature $(2, 2)$ (we refer to García-Río and Vázquez-Lorenzo [92]).

THE SKEW-SYMMETRIC CURVATURE OPERATOR

Theorem 2.6 (1) generalizes to this setting to give an appropriate characterization of affine Ivanov–Petrova surfaces in terms of the Ricci tensor.

Theorem 2.7   *Let (Σ, ∇) be a connected 2-dimensional affine surface.*

1. (Σ, ∇) is affine Ivanov–Petrova at P if and only if ρ is symmetric and degenerate.

2. Let (Σ, ∇) have recurrent curvature.

(a) (Σ, ∇) is affine Ivanov–Petrova if and only if there exist coordinate systems centered at an arbitrary point $P \in \Sigma$ so that the only (possibly) non-zero Christoffel symbol is $\Gamma_{11}{}^2 = a(x^1, x^2)$.

(b) If (Σ, ∇) is affine Ivanov–Petrova, then (Σ, ∇) is locally symmetric if and only if $a(x^1, x^2) = \alpha\, x^2 + \xi(x^1)$ for $\alpha \in \mathbb{R}$. In this setting, (Σ, ∇) is projectively flat; (Σ, ∇) is flat if and only if $\alpha = 0$.

Proof. Let $\{e_1, e_2\}$ be linearly independent vectors in $T_P \Sigma$; since the dimension is two, they form a basis. We apply Equation (2.3.a) to see that $\mathcal{R}$ is nilpotent if and only if $\rho_{12} - \rho_{21} = 0$ (or, equivalently, if ρ is symmetric) and $\det\{\rho\} = 0$ (or, equivalently, if ρ is degenerate). Assertion (1) follows.

    We use Assertion (1) to see that (Σ, ∇) is affine Ivanov–Petrova if and only if $\rho_a = 0$ and $\det\{\rho_s\} = 0$. Since the curvature is assumed to be recurrent, the work of Wong [213] shows there exist local coordinates (x^1, x^2) so that $\Gamma_{11}{}^2 = a(x^1, x^2)\partial_{x_2}$ and so the other Christoffel symbols vanish; the assumption (Σ, ∇) is not flat then yields $\partial_{x_2} a(x^1, x^2) \neq 0$. We check that the only non-vanishing component of the Ricci tensor is $\rho(\partial_{x_1}, \partial_{x_1}) = \partial_{x_2} a(x^1, x^2)$ and (Σ, ∇) is locally symmetric if and only if $a(x^1, x^2) = \alpha\, x^2 + \xi(x^1)$; Assertion (2) now follows. $\square$

2.4   HOMOGENEOUS 2-DIMENSIONAL AFFINE SURFACES

An affine manifold (M, ∇) is said to be *locally homogeneous* if, for any two points $P, Q \in M$, there exists a local diffeomorphism φ from a neighborhood of P to a neighborhood of Q so that $\varphi^* \nabla = \nabla$. We refer to Opozda [172] for the proof of Assertion (2) and to Arias-Marco and Kowalski [9] for the proof of Assertions (1,3,4) in the following result:

Theorem 2.8   *Let (Σ, ∇) be an affine surface.*

1. *(Σ, ∇) is locally homogeneous if and only if it admits, in a neighborhood of each point of Σ, at least two linearly independent affine-Killing vector fields.*

2. *If (Σ, ∇) is locally homogeneous, then either ∇ is the Levi-Civita connection of a metric of constant curvature, or there exist local coordinates (x^1, x^2) and suitably chosen constants $\{a, b, c, d, e, f\}$ so that the connection has one of the following forms:*

   Type A: $\nabla_{\partial_{x_1}} \partial_{x_1} = a\partial_{x_1} + b\partial_{x_2}, \; \nabla_{\partial_{x_1}} \partial_{x_2} = c\partial_{x_1} + d\partial_{x_2}, \; \nabla_{\partial_{x_2}} \partial_{x_2} = e\partial_{x_1} + f\partial_{x_2}.$

   Type B: $\nabla_{\partial_{x_1}} \partial_{x_1} = \dfrac{a\partial_{x_1} + b\partial_{x_2}}{x^1}, \; \nabla_{\partial_{x_1}} \partial_{x_2} = \dfrac{c\partial_{x_1} + d\partial_{x_2}}{x^1}, \; \nabla_{\partial_{x_2}} \partial_{x_2} = \dfrac{e\partial_{x_1} + f\partial_{x_2}}{x^1}.$

3. *(Σ, ∇) is of Type A if and only if it locally admits a pair of linearly independent affine-Killing vector fields X, Y such that $[X, Y] = 0$.*

4. (Σ, ∇) *is of Type B if and only if it locally admits a pair of linearly independent affine-Killing vector fields X, Y such that $[X, Y] = X$.*

We emphasize that there are connections which are both Type A and Type B; we shall discuss the intersection of these classes presently in Theorem 2.12 and in Theorem 2.13. We must first examine some geometric properties of Type A and Type B locally homogeneous connections. Of particular interest are those that are projectively flat and recurrent. Recall that an equiaffine surface is *projectively flat* if and only if the Ricci tensor is a *Codazzi tensor*, or, equivalently, if we have that $(\nabla_X \rho)(Y, Z) = (\nabla_Y \rho)(X, Z)$.

TYPE A LOCALLY HOMOGENEOUS AFFINE SURFACES

Recall that we have the local coordinate representation:

$$\nabla_{\partial_{x_1}} \partial_{x_1} = a\partial_{x_1} + b\partial_{x_2}, \ \nabla_{\partial_{x_1}} \partial_{x_2} = c\partial_{x_1} + d\partial_{x_2}, \ \nabla_{\partial_{x_2}} \partial_{x_2} = e\partial_{x_1} + f\partial_{x_2}.$$

We summarize the geometric properties of such a connection as follows:

Theorem 2.9   *Let (Σ, ∇) be a Type A locally homogeneous affine surface. Then the Ricci tensor is symmetric and given by:*

$$\rho_{11} = -d^2 + ad + (f - c)b, \ \rho_{12} = \rho_{21} = cd - eb, \ \rho_{22} = -c^2 + fc + (a - d)e.$$

Either the Ricci tensor defines a flat metric on Σ or (Σ, ∇) is affine Ivanov–Petrova. The covariant derivatives of the Ricci tensor take the form:

$$\tfrac{1}{2}(\nabla_{\partial_{x_1}} \rho)(\partial_{x_1}, \partial_{x_1}) = -da^2 + (d^2 - bf + cb)a + (be - cd)b,$$

$$\tfrac{1}{2}(\nabla_{\partial_{x_2}} \rho)(\partial_{x_1}, \partial_{x_1}) = \tfrac{1}{2}(\nabla_{\partial_{x_1}} \rho)(\partial_{x_1}, \partial_{x_2}) = -acd + (c^2 - fc + de)b,$$

$$\tfrac{1}{2}(\nabla_{\partial_{x_2}} \rho)(\partial_{x_1}, \partial_{x_2}) = \tfrac{1}{2}(\nabla_{\partial_{x_1}} \rho)(\partial_{x_2}, \partial_{x_2}) = bce - (ae + cf - de)d,$$

$$\tfrac{1}{2}(\nabla_{\partial_{x_2}} \rho)(\partial_{x_2}, \partial_{x_2}) = fc^2 - (de + f^2)c - (af - be - df)e.$$

Consequently, (Σ, ∇) is projectively flat. Finally, (Σ, ∇) is affine Ivanov–Petrova if and only if it is recurrent.

Proof. The computation of ρ and of $\nabla\rho$ follows directly from the defining property; the fact that either ρ defines a flat metric or (Σ, ∇) is affine Ivanov–Petrova now follows, as does the fact that (Σ, ∇) is projectively flat. We argue as follows to establish the remaining assertion. Suppose first that (Σ, ∇) is affine Ivanov–Petrova or, equivalently, by Theorem 2.7, that the Ricci tensor is symmetric and degenerate. Note that ρ is degenerate if and only if

$$b^2 e^2 - \left\{ d^3 - 2ad^2 + (a^2 + 3bc - bf)d + (f - c)ab \right\} e$$

$$+ \left\{ fd^2 + a(c - f)d - b(c - f)^2 \right\} c = 0. \tag{2.4.a}$$

We examine the possibilities seriatim to show (Σ, ∇) is recurrent. We will show either that (Σ, ∇) is locally symmetric (in which case $\nabla\rho = 0$) or we will construct ω so $\nabla\rho = \omega \otimes \rho$.

1. If $b = 0$, then we obtain several sub-cases since Equation (2.4.a) reduces to

$$d \cdot \left\{ ac^2 - (a-d)fc - (a-d)^2 e \right\} = 0. \tag{2.4.b}$$

(a) If $d = 0$ then $\nabla\rho = \omega \otimes \rho$, with $\omega = (-2f)dx^2$.

(b) If $d \neq 0$ and if $a = 0$, then $e = d^{-1}cf$ and $\nabla\rho = 0$.

(c) If $d \neq 0$ and if $a \neq 0$, then Equation (2.4.b) shows that

$$ac^2 - (a-d)fc - (a-d)^2 e = 0, \quad \text{so,}$$
$$c = (2a)^{-1}(a-d)(f + \varepsilon(f^2 + 4ae)^{\frac{1}{2}}),$$

with $\varepsilon \in \{-1, 1\}$ (c being real). If $d = a$, then the connection is symmetric and $\nabla\rho = 0$ so (Σ, ∇) is recurrent. On the other hand, if $d \neq a$, then $\nabla\rho = \omega \otimes \rho$, where $\omega = (-2a)dx^1 - (f + \varepsilon(f^2 + 4ae)^{\frac{1}{2}})dx^2$.

2. If $b \neq 0$, then Equation (2.4.a) yields:

$$e = (2b^2)^{-1}\left\{ d^3 - 2ad^2 + (a^2 + 3bc - bf)d + (f - c)ab + \varepsilon(d^2 - ad + (c - f)b)\zeta^{\frac{1}{2}} \right\},$$

with $\zeta = (a-d)^2 + 4bc$ and $\varepsilon \in \{-1, 1\}$, as e is real. Now, if $c = b^{-1}(-d^2 + ad + bf)$, then $\nabla\rho = 0$. Otherwise, $\nabla\rho = \omega \otimes \rho$ with

$$\omega = -(a + d + \varepsilon\zeta^{\frac{1}{2}})dx^1 - b^{-1}(d^2 - ad + 2bc + \varepsilon d\zeta^{\frac{1}{2}})dx^2.$$

Conversely, suppose that (Σ, ∇) is recurrent. First, observe that, if ∇ is locally symmetric, then the vanishing of the covariant derivative of the Ricci tensor shows that Equation (2.4.a) vanishes. Hence, the Ricci tensor is symmetric and degenerate and thus, (Σ, ∇) is affine Ivanov–Petrova. Next, if (Σ, ∇) is recurrent but non-symmetric, then there exists a 1-form $\omega = \omega_1 dx^1 + \omega_2 dx^2$ such that $\nabla\rho = \omega \otimes \rho$, where at least one of ω_1 or ω_2 is non-zero. Since any Type A connection is projectively flat, one has that

$$\omega_1 \rho_{22} = \omega_2 \rho_{12} \quad \text{and} \quad \omega_2 \rho_{11} = \omega_1 \rho_{12}.$$

We may assume that $\omega_1 \neq 0$, as the case $\omega_2 \neq 0$ is analogous. Identities derived previously permit us to conclude that (Σ, ∇) is affine Ivanov–Petrova since

$$\det \rho = \rho_{11}\rho_{22} - \rho_{12}^2 = \tfrac{\omega_2}{\omega_1}\rho_{11}\rho_{12} - \rho_{12}^2 = 0. \qquad \square$$

TYPE B LOCALLY HOMOGENEOUS AFFINE SURFACES

Recall that we have the local coordinate representation:

$$\nabla_{\partial_{x_1}} \partial_{x_1} = \frac{a\partial_{x_1} + b\partial_{x_2}}{x^1}, \quad \nabla_{\partial_{x_1}} \partial_{x_2} = \frac{c\partial_{x_1} + d\partial_{x_2}}{x^1}, \quad \nabla_{\partial_{x_2}} \partial_{x_2} = \frac{e\partial_{x_1} + f\partial_{x_2}}{x^1}.$$

We now summarize the relevant facts we shall need:

Theorem 2.10   *Let (Σ, ∇) be a Type B locally homogeneous equiaffine surface. Then*

$$\rho_{11} = \tfrac{1}{(x^1)^2}\{(a - d + 1)d + (f - c)b\}, \quad \rho_{12} = \tfrac{1}{(x^1)^2}\{cd - be + f\},$$
$$\rho_{21} = \tfrac{1}{(x^1)^2}\{cd - be - c\}, \quad \rho_{22} = \tfrac{1}{(x^1)^2}\{(a - d - 1)e + (f - c)c\}.$$

Either ρ_s defines a metric of constant Gauss curvature on Σ, or (Σ, ∇) is affine Ivanov–Petrova. Moreover,

$$\tfrac{1}{2}(x^1)^3(\nabla_{\partial_{x_1}} \rho)(\partial_{x_1}, \partial_{x_1}) = (a + 1)(d - a - 1)d + (2a - d + 3)bc + b^2 e,$$
$$\tfrac{1}{2}(x^1)^3(\nabla_{\partial_{x_2}} \rho)(\partial_{x_1}, \partial_{x_1}) = 2bc^2 - adc + bde,$$
$$(x^1)^3(\nabla_{\partial_{x_1}} \rho)(\partial_{x_1}, \partial_{x_2}) = (a + 4bc - 2(a + 1)d + 2)c + (2d + 3)be,$$
$$\tfrac{1}{2}(x^1)^3(\nabla_{\partial_{x_2}} \rho)(\partial_{x_1}, \partial_{x_2}) = c^2 d + bec + (d - a)de,$$
$$\tfrac{1}{2}(x^1)^3(\nabla_{\partial_{x_1}} \rho)(\partial_{x_2}, \partial_{x_2}) = (d + 1)(d - a + 1)e + (d + 3)c^2 + bce,$$
$$\tfrac{1}{2}(x^1)^3(\nabla_{\partial_{x_2}} \rho)(\partial_{x_2}, \partial_{x_2}) = -2c^3 + be^2 + (a - 2d)ce.$$

In this setting (Σ, ∇) is projectively flat if and only if

$$\{e = f = c = 0\} \text{ or } \left\{e \neq 0, f = -c, a = \tfrac{3c^2 + 2de + e}{e}, b = \tfrac{-c^3 - ce}{e^2}\right\}.$$

Proof. One computes the Ricci tensor directly; this shows that ρ need not be symmetric and thus, ∇ need not be equiaffine; ρ is symmetric if and only if $f = -c$. Suppose that ρ_s defines a metric on Σ, then the associated Gauss curvature satisfies

$$\begin{aligned}
K &= -(x^1)^{-2}\rho_{22} \det\{\rho\}^{-1}\\
&= \left(2c^2 + e(d - a + 1)\right)\\
&\quad \times \left(c(4bc^2 + (d^2 - 2ad - 1)c + 2(2d - a)be) + e(((a - d)^2 - 1)d - b^2 e)\right)^{-1}
\end{aligned}$$

and K is constant. One computes $\nabla\rho$ directly. It then follows that (Σ, ∇) is projectively flat if and only if

$$0 = (\nabla_{\partial_{x_1}} \rho)(\partial_{x_2}, \partial_{x_1}) - (\nabla_{\partial_{x_2}} \rho)(\partial_{x_1}, \partial_{x_1}) = \frac{c(a - 2d + 2) + 3be}{(x^1)^3},$$
$$0 = (\nabla_{\partial_{x_1}} \rho)(\partial_{x_2}, \partial_{x_2}) - (\nabla_{\partial_{x_2}} \rho)(\partial_{x_1}, \partial_{x_2}) = \frac{2(3c^2 - ae + 2de + e)}{(x^1)^3}.$$

The final implication now follows.   □

We note that, in general, an affine manifold (M, ∇) is projectively flat if and only if (T^*M, g_∇) is locally conformally flat, which is equivalent to the vanishing of the associated Weyl conformal curvature tensor (see, for example, Afifi [3]). Hence, if a Type B locally homogeneous affine surface is projectively flat, then the associated Ricci tensor is necessarily symmetric, since $W(\partial_{x_1}, \partial_{x_2}, \partial_{x_1}, \partial_{x^{1\prime}}) = \frac{c+f}{2(x^1)^2}$, which shows that $f = -c$, the necessary and sufficient condition for ρ to be symmetric.

Analyzing which affine connections are recurrent is more involved, but it shows a clear link with the Ivanov–Petrova property. First of all, note that Type A locally homogeneous affine connections need not be recurrent in general. Set $b = c = 0$ in Theorem 2.8 to get $\rho_{12} = 0$, but $(\nabla_{\partial_{x_2}} \rho)(\partial_{x_1}, \partial_{x_2}) = 2de(d-a)$, which shows that the connection is not recurrent in general.

We now determine which Type B locally homogeneous affine surfaces are Ivanov–Petrova. Unlike the Type A setting, these surfaces are not necessarily recurrent.

Theorem 2.11   *Let (Σ, ∇) be a Type B locally homogeneous affine surface. Then (Σ, ∇) is affine Ivanov–Petrova if and only if either (Σ, ∇) is recurrent or if the Christoffel symbols in Theorem 2.8 are given by*

$$\nabla_{\partial_{x_1}} \partial_{x_1} = \tfrac{a}{x^1}\partial_{x_1} + \tfrac{b}{x^1}\partial_{x_2}, \quad \nabla_{\partial_{x_1}} \partial_{x_2} = \tfrac{c}{x^1}\partial_{x_1} + \tfrac{d}{x^1}\partial_{x_2}, \quad \nabla_{\partial_{x_2}} \partial_{x_2} = \tfrac{e}{x^1}\partial_{x_1} - \tfrac{c}{x^1}\partial_{x_2},$$

for real constants a, b, c, d, and e belonging to one of the following eight cases:

1. $b = 0, d \neq 0, e = 0, c \neq 0, a = (2d)^{-1}(d^2 - 1)$.

2. $b = 0, de \neq 0, c = 0, a = d \pm 1$.

3. $b = 0, de \neq 0, c \neq 0, e \neq -d^{-1}c^2, a = (de)^{-1}(d(c^2 + de) \pm \zeta^{\frac{1}{2}})$, with

   $\zeta = d(c^2d + e)(c^2 + de) \geq 0$.

4. $b \neq 0, d = 0, c \neq 0, e = (b^2)^{-1}(-abc \pm (b^2c^2(a^2 + 4bc - 1))^{\frac{1}{2}})$ with $a^2 + 4bc - 1 \geq 0$.

5. $b \neq 0, d \neq 0, a \neq \pm(d-1), c = (2b)^{-1}(a-d+1)d, e = (2b^2)^{-1}(a-d+1)(d-1)d$.

6. $b \neq 0, d \neq 0, c \notin \{0, b^{-1}ad, (2b)^{-1}(a-d+1)d\}$,

   $e = (2b^2)^{-1}((d-a)^2 + 4bc - 1)d - 2abc \pm \zeta^{\frac{1}{2}}$,

   with $\zeta = ((d-a+1)d + 2bc)((d-a-1)d + 2bc)((d-a)^2 + 4bc - 1) \geq 0$.

7. $b \neq 0, d \neq 0, c = 0, a \neq d - 1$,

   $e = (2b^2)^{-1}(d^3 - 2ad^2 + (a^2 - 1)d \pm |d||(a-d)^2 - 1|) \neq 0$.

8. $b \neq 0, d \neq 0, c = \frac{ad}{b}, a \neq 1 - d$,

$$e = (2b^2)^{-1}(d^3 + 2ad^2 - (a^2 + 1)d \pm |d||(a + d)^2 - 1|) \neq -(b^2)^{-1}(a^2d).$$

Proof. Assume that (Σ, ∇) is affine Ivanov–Petrova. We suppose $f = -c$ henceforth. Theorem 2.10 implies that ρ is symmetric if and only if $f = -c$. Thus, ρ is degenerate if and only if

$$b^2e^2 - \left\{ d^3 - 2ad^2 + (a^2 + 4bc - 1)d - 2abc \right\} e - \left\{ d^2 - 2ad + 4bc - 1 \right\} c^2 = 0. \quad (2.4.c)$$

We assume that $\nabla\rho = \omega_1 dx^1 + \omega_2 dx^2$ and examine the implications of this identity. We proceed as in the proof Theorem 2.9 using Theorem 2.10 to examine the solutions of Equation (2.4.c). We first observe that if $b = 0$, then Equation (2.4.c) becomes:

$$dea^2 - 2d(c^2 + de)a + (d^2 - 1)(c^2 + de) = 0.$$

This leads to five cases which we examine seriatim:

1. Suppose $d = 0$. Then $c = 0$ so $\nabla\rho = \omega \otimes \rho$ with $\omega = -\frac{2}{x^1}dx^1$.

2. Suppose $d \neq 0$, $e = 0$, and $c = 0$. Then $\nabla\rho = \omega \otimes \rho$, with $\omega = -\frac{2+2a}{x^1}dx^1$.

3. Suppose $d \neq 0$, $e = 0$, and $c \neq 0$. Then $a = (2d)^{-1}(d^2 - 1)$ and (Σ, ∇) is not recurrent—this leads to Case 1. If $\nabla\rho = \omega \otimes \rho$, then

$$(\nabla_{\partial_{x_2}}\rho)(\partial_{x_2}, \partial_{x_2}) - \omega_2\rho_{22} = \frac{2c^2(x^1\omega_2 - 2c)}{(x^1)^3},$$

so $\omega_2 = \frac{2c}{x^1}$ and $(\nabla_{\partial_{x_2}}\rho)(\partial_{x_1}, \partial_{x_2}) - \omega_2\rho_{12} = \frac{2c^2}{(x^1)^3} \neq 0$ and ∇ is not recurrent.

4. Suppose $d \neq 0$ and $0 \neq e$. Then $a = (de)^{-1}(d(c^2 + de) + \varepsilon\zeta^{\frac{1}{2}})$ with $\varepsilon = \pm 1$ and with $\zeta = d(c^2d + e)(c^2 + de) \geq 0$. We suppose $\nabla\rho = \omega \otimes \rho$, where $\omega = \omega_1 dx^1 + \omega_2 dx^2$ and examine the implications. If $c = 0$, then $\zeta = d^2e^2 \geq 0$ and $a = d \pm 1$. Consequently,

$$0 = (\nabla_{\partial_{x_2}}\rho)(\partial_{x_1}, \partial_{x_2}) - \omega_2\rho_{12} = -2\varepsilon|d||e|(x^1)^{-3} \neq 0,$$

which gives Case 2. If $c \neq 0$, then

$$\begin{aligned}
0 &= (x^1)^3\left((\nabla_{\partial_{x_1}}\rho)(\partial_{x_1}, \partial_{x_2}) - \omega_1\rho_{12}\right) \\
&= \frac{c^3d(1-2d) - cde(2d^2 + d - 2) - \varepsilon c(2d-1)\zeta^{\frac{1}{2}}}{de} - c(d-1)x^1\omega_1, \quad (2.4.d) \\
0 &= (x^1)^3\left((\nabla_{\partial_{x_2}}\rho)(\partial_{x_1}, \partial_{x_2}) - \omega_2\rho_{12}\right) = -2\varepsilon\zeta^{\frac{1}{2}} - c(d-1)x^1\omega_2.
\end{aligned}$$

Assuming $d = 1$, the second equation above reduces to

$$(\nabla_{\partial_{x_2}}\rho)(\partial_{x_1}, \partial_{x_2}) - \omega_2\rho_{12} = \frac{-2\varepsilon|c^2 + e|}{(x^1)^3},$$

hence, ∇ is not recurrent if $e \neq -c^2$, thus showing Case 3 (for $d = 1$). Note that, for $e = -c^2$, we have $\nabla\rho = 0$. Next, if $d \neq 1$, then ω_1 and ω_2 are determined by Equation (2.4.d) as follows

$$d(x^1)^3\left((\nabla_{\partial_{x_1}}\rho)(\partial_{x_2},\partial_{x_2})-\omega_1\rho_{22}\right)=c^2+de-\varepsilon\zeta^{\frac{1}{2}},$$

$$\tfrac{1}{2}c(d-1)e(x^1)^3\left((\nabla_{\partial_{x_2}}\rho)(\partial_{x_1},\partial_{x_1})-\omega_2\rho_{11}\right)=(c^2+de)(d(c^2+e)+\varepsilon\zeta^{\frac{1}{2}}),$$

$$\tfrac{1}{2}c(d-1)d(x^1)^3\left((\nabla_{\partial_{x_2}}\rho)(\partial_{x_2},\partial_{x_2})-\omega_2\rho_{22}\right)=(c^2+de)(d(c^2+e)-\varepsilon\zeta^{\frac{1}{2}}).$$

Note that the three expressions above do not vanish simultaneously for $e\neq-\frac{c^2}{d}$ and hence, in such a case, the affine connection is not recurrent (Case 3 with $d\neq1$). If $e=d^{-1}c^2$ then, setting $\omega=-\frac{2}{x^1}dx^1$, it follows that $\nabla\rho=\omega\otimes\rho$ holds.

5. Suppose $b\neq0$. Let $\varepsilon=\pm1$. By Equation (2.4.c), we have that:

$$\begin{aligned}e&=(2b^2)^{-1}\left\{((d-a)^2+4bc-1)d-2abc+\varepsilon\zeta^{\frac{1}{2}}\right\}\text{ with}\\\zeta&=((d-a+1)d+2bc)((d-a-1)d+2bc)((d-a)^2+4bc-1)\geq0.\end{aligned}$$

Assume $\nabla\rho=\omega\otimes\rho$. If $d=0$, then

$$(2b)^{-1}(x^1)^3\left((\nabla_{\partial_{x_2}}\rho)(\partial_{x_1},\partial_{x_1})-\omega_2\rho_{11}\right)=c(2c+x^1\omega_2).$$

Hence, for $c\neq0$, it follows that $\omega_2=-\frac{2c}{x^1}$ and we get

$$(\nabla_{\partial_{x_2}}\rho)(\partial_{x_1},\partial_{x_2})-\omega_2\rho_{12}=-(x^1)^{-3}2c^2,$$

which does not vanish (Case 4). If $d\neq0$, then:

$$(x^1)^3\left((\nabla_{\partial_{x_1}}\rho)(\partial_{x_1},\partial_{x_1})-\omega_1\rho_{11}\right)$$

$$=(a+d+3)((d-a-1)d+2bc)+\varepsilon\zeta^{\frac{1}{2}}+((d-a-1)d+2bc)x^1\omega_1,$$

$$(x^1)^3\left((\nabla_{\partial_{x_2}}\rho)(\partial_{x_1},\partial_{x_1})-\omega_2\rho_{11}\right)$$

$$=b^{-1}\left\{((d-a-1)d+2bc)((d-a+1)d+2bc)+d\varepsilon\zeta^{\frac{1}{2}}\right\}$$

$$+((d-a-1)d+2bc)x^1\omega_2.$$

Now, if $(d-a-1)d+2bc=0$ or, equivalently, if $c=(2b)^{-1}(a-d+1)d$, then the expressions above vanish and

$$(\nabla_{\partial_{x_1}}\rho)(\partial_{x_1},\partial_{x_2})-\omega_1\rho_{12}=(2b(x^1)^3)^{-1}(a^2-(d-1)^2)d.$$

It follows then that, for $a \neq \pm(d-1)$, the affine connection is not recurrent (Case 5), while for $a = \pm(d-1)$, one has $\nabla \rho = 0$. On the other hand, if $c \neq (2b)^{-1}(a-d+1)d$, then ω_1 and ω_2 are determined by the expressions above and one may compute that:

$$(2b)^{-1}d((\nabla_{\partial_{x_1}} \rho)(\partial_{x_1}, \partial_{x_2}) - \omega_1 \rho_{12}) - ((\nabla_{\partial_{x_2}} \rho)(\partial_{x_1}, \partial_{x_2}) - \omega_2 \rho_{12})$$
$$= (b(x^1)^3)^{-1}2c(bc - ad).$$

Therefore, if $c(bc - ad) \neq 0$ then the affine connection is not recurrent (Case 6). If we have $c = 0$, then $c \neq (2b)^{-1}(a-d+1)d$. This implies that $a \neq d-1$. Furthermore, $\nabla \rho = \omega \otimes \rho$ if and only if $e = 0$ (Case 7). If $bc - ad = 0$, then $c \neq (2b)^{-1}(a-d+1)d$. Thus, $a \neq 1-d$. This shows the connection is recurrent if and only if $e = -(b^2)^{-1}a^2 d$, which yields Case 8.

Conversely, any equiaffine homogeneous connection of Type B that is recurrent, or given by the conditions above, is affine Ivanov–Petrova (see Calviño-Louzao, García-Río, and Vázquez-Lorenzo [45]). □

AFFINE SURFACES WHICH ARE BOTH TYPE A AND TYPE B

Theorem 2.12 *A non-flat Type B affine connection is affinely equivalent to a Type A affine connection if and only if $\nabla_{\partial_{x_1}} \partial_{x_1} = \frac{1}{x^1}(a\partial_{x_1} + b\partial_{x_2})$, $\nabla_{\partial_{x_1}} \partial_{x_2} = \frac{1}{x^1}d\partial_{x_2}$, and $\nabla_{\partial_{x_2}} \partial_{x_2} = 0$.*

Proof. We apply Theorem 2.9 and Theorem 2.10. First of all, recall that any Type A affine connection is projectively flat. Moreover, the Ricci tensor is always symmetric and defines a flat metric on the surface or, otherwise, it is degenerate. Further note that, in the later case, the Ricci tensor is always recurrent. Observe that the projectively flat Type B locally homogeneous connections are listed in Theorem 2.10:

$$e = f = c = 0 \tag{2.4.e}$$
$$e \neq 0, \quad f = -c, \quad a = \tfrac{3c^2+2de+e}{e}, \quad b = \tfrac{-c^3-ce}{e^2}. \tag{2.4.f}$$

In the situation given in Equation (2.4.f), if the Ricci tensor is non-degenerate, then it defines a metric of non-zero curvature and thus, no such connection can be affinely equivalent to any of Type A. Moreover, if the Ricci tensor is degenerate but non-zero, then we show that such a connection cannot be projectively flat and recurrent simultaneously, and hence, they cannot be affinely equivalent to any Type A connection.

First assume that Equation (2.4.f) pertains with $c = 0$. Then the Ricci tensor is degenerate for $d = 0$ or $d = -2$. In the case $d = 0$, the connection is flat, while, if $d = -2$, the connection is not recurrent. Next assume that Equation (2.4.f) holds with $c \neq 0$. In this case, the Ricci tensor is degenerate if and only if $d = -\frac{c^2}{e}$ or $d = -\frac{c^2}{e} - 2$. If $d = -\frac{c^2}{e}$, then the connection is flat. Hence, set $d = -\frac{c^2}{e} - 2$. If $b = 0$, we have the conditions $a = -4, c \neq 0, d = -1$ and $e = -c^2$. This shows that the connection is not recurrent. Thus, in any case the connection is not recurrent.

Next we show that any connection of Type B given by Equation (2.4.e) is affinely equivalent to a Type A connection. In doing that, we recall the discussion in Arias-Marco and Kowalski [9] and in Kowalski, Opozda, and Vlášek [144]: Type A locally homogeneous affine connections are characterized by the existence of linearly independent commuting affine-Killing vector fields.

Recall that a vector field X is said to be *affine-Killing* if

$$[X, \nabla_Y Z] - \nabla_Y [X, Z] - \nabla_{[X,Y]} Z = 0 \text{ for all } Y, Z.$$

Thus, a vector field $X = A(x^1, x^2)\partial_{x_1} + B(x^1, x^2)\partial_{x_2}$ on the coordinate domain $\mathcal{U}(x^1, x^2)$ is affine-Killing for the connection in Equation (2.4.e) if and only if

$$A_{11} + \frac{a}{x^1}A_1 - \frac{b}{x^1}A_2 - \frac{a}{(x^1)^2}A = 0, \qquad A_{12} + \frac{a-d}{x^1}A_2 = 0, \qquad A_{22} = 0,$$

$$B_{11} + \frac{2b}{x^1}A_1 + \frac{2d-a}{x^1}B_1 - \frac{b}{x^1}B_2 - \frac{b}{(x^1)^2}A = 0,$$

$$B_{12} + \frac{d}{x^1}A_1 + \frac{b}{x^1}A_2 - \frac{d}{(x^1)^2}A = 0, \qquad B_{22} + \frac{2d}{x^1}A_2 = 0,$$

where the subscripts mean $A_{ij} = \partial_{x_i}\partial_{x_j}A(x^1, x^2)$ (similarly for the function B). Set

$$X = \left\{ \begin{array}{ll} x^1\partial_{x_2} & \text{if } a - 2d = 0 \\ \partial_{x_2} & \text{if } a - 2d \neq 0 \end{array} \right\} \text{ and}$$

$$Y = \left\{ \begin{array}{ll} x^1\partial_{x_1} + (x^1 + x^2)\partial_{x_2} & \text{if } a - 2d = 0 \\ x^1\partial_{x_1} + \frac{b}{a-2d}x^1\partial_{x_2} & \text{if } a - 2d \neq 0 \end{array} \right\}.$$

Then $\{X, Y\}$ are linearly independent commuting affine-Killing vector fields, and hence, the connection is of Type A. □

Observe the results established above show that the locally homogeneous connections given in Theorem 2.12 are projectively flat and recurrent with symmetric and degenerate Ricci tensor. However, not every projectively flat and recurrent locally homogeneous affine connection with symmetric and degenerate Ricci tensor is necessarily of Type B, as the following example shows; we refer to Calviño-Louzao, García-Río, and Vázquez-Lorenzo [45] for further details.

Theorem 2.13   *Let (Σ, ∇) be an affine surface with symmetric and degenerate Ricci tensor, which is recurrent and projectively flat. Then (Σ, ∇) is locally homogeneous if and only if around each point there exists a coordinate system (x^1, x^2) in which the non-zero component of ∇ is given by $\nabla_{\partial_{x_1}}\partial_{x_1} = x^2\frac{\mu}{(\alpha+\kappa x^1)^2}\partial_{x_2}$ for some constants μ, α and κ. Moreover, any such a connection is locally homogeneous of both Type A and Type B if and only if $\kappa^2 - 4\mu \geq 0$.*

A connection of the form given in Theorem 2.13 is flat if and only if $\mu = 0$ and locally symmetric if and only if $\kappa = 0$. Hence, a projectively flat locally symmetric connection with symmetric and degenerate Ricci tensor is of Type B if and only if the only non-zero Christoffel symbol is $\Gamma_{11}^2 = Kx^2$ with $K \leq 0$.

2.5 THE SPECTRAL GEOMETRY OF THE CURVATURE TENSOR OF DEFORMED RIEMANNIAN EXTENSIONS

Lemma 2.14 *Let (M, ∇) be an affine manifold, let $\tilde{\nabla}_{\nabla,\Phi}$ be the Levi-Civita connection of the deformed Riemannian extension $(T^*M, g_{\nabla,\Phi})$, and let $\sigma : T^*M \to M$ be the canonical projection. Let $\{\tilde{X}, \tilde{Y}\}$ be tangent vectors on T^*M and let $\{X := \sigma_*\tilde{X}, Y := \sigma_*\tilde{Y}\}$ be the corresponding tangent vectors on M. Then*

$$\mathrm{Spec}\{\tilde{\mathcal{J}}(\tilde{X})\} = \mathrm{Spec}\{\mathcal{J}(X)\} \text{ and}$$
$$\mathrm{Spec}\{\tilde{\mathcal{R}}(\tilde{X}, \tilde{Y})\} = \mathrm{Spec}\{\mathcal{R}(X, Y)\} \cup -\mathrm{Spec}\{\mathcal{R}(X, Y)\}.$$

Proof. Let $\Gamma_{ij}{}^k$ be the Christoffel symbols of ∇ and let Φ_{ij} be the components of Φ. We use Lemma 1.31 to see that the (possibly) non-zero Christoffel symbols $\tilde{\Gamma}_{\alpha\beta}{}^\gamma$ of the Levi-Civita connection of $g_{\nabla,\Phi}$ are given by:

$$\tilde{\Gamma}_{ij}{}^k = \Gamma_{ij}{}^k, \qquad \tilde{\Gamma}_{i'j}{}^{k'} = -\Gamma_{jk}{}^i, \qquad \tilde{\Gamma}_{ij'}{}^{k'} = -\Gamma_{ik}{}^j,$$

$$\tilde{\Gamma}_{ij}{}^{k'} = \sum_r x_{r'}\left(\partial_{x_k}\Gamma_{ij}{}^r - \partial_{x_i}\Gamma_{jk}{}^r - \partial_{x_j}\Gamma_{ik}{}^r + 2\sum_\ell \Gamma_{k\ell}{}^r \Gamma_{ij}{}^\ell\right)$$

$$+ \frac{1}{2}\left(\partial_{x_i}\Phi_{jk} + \partial_{x_j}\Phi_{ik} - \partial_{x_k}\Phi_{ij}\right) - \sum_\ell \Phi_{k\ell}\Gamma_{ij}{}^\ell.$$

By Lemma 1.31, the non-zero components of the curvature tensor of $(T^*M, g_{\nabla,\Phi})$ up to the usual symmetries are given as follows; we omit $\tilde{R}_{kji}{}^{h'}$, as it plays no role in our considerations:

$$\tilde{R}_{kji}{}^h = R_{kji}{}^h, \qquad \tilde{R}_{kji}{}^{h'}, \qquad \tilde{R}_{kji'}{}^{h'} = -R_{kjh}{}^i, \qquad \tilde{R}_{k'ji}{}^{h'} = R_{hij}{}^k.$$

Let $\tilde{X} = \alpha^i \partial_{x_i} + \alpha_{i'}\partial_{x_{i'}}$ and $\tilde{Y} = \beta^i\partial_{x_i} + \beta_{i'}\partial_{x_{i'}}$ be vector fields on T^*M. Let $X = \alpha^i\partial_{x_i}$ and $Y = \beta^i\partial_{x_i}$ be the corresponding vector fields on M. Let $\mathcal{J}(X)$ and $\mathcal{R}(X, Y)$ be the matrices of the Jacobi operator and of the skew-symmetric curvature operator on M relative to the basis $\{\partial_{x_i}\}$. Then the matrix of the Jacobi operator $\tilde{\mathcal{J}}(\tilde{X})$ and of the skew-symmetric curvature operator $\tilde{\mathcal{R}}(\tilde{X}, \tilde{Y})$ with respect to the basis $\{\partial_{x_i}, \partial_{x_{i'}}\}$ have the form:

$$\tilde{\mathcal{J}}(\tilde{X}) = \begin{pmatrix} \mathcal{J}(X) & 0 \\ * & {}^t\mathcal{J}(X) \end{pmatrix} \text{ and } \tilde{\mathcal{R}}(\tilde{X}, \tilde{Y}) = \begin{pmatrix} \mathcal{R}(X, Y) & 0 \\ * & -{}^t\mathcal{R}(X, Y) \end{pmatrix}.$$

The desired result now follows. $\qquad\square$

The following is the main result of this section:

Theorem 2.15

1. Let (M, g) be a 4-dimensional self-dual Walker manifold.

   (a) If (M, g) is Ricci flat, then (M, g) is locally isometric to a deformed Riemannian extension.

   (b) If (M, g) is Ivanov–Petrova, then (M, g) is locally isometric to a deformed Riemannian extension.

2. Let (M, ∇) be an affine manifold. Let $\tilde{\nabla}_{\nabla,\Phi}$ be the Levi-Civita connection of $(T^*M, g_{\nabla,\Phi})$.

   (a) The following Assertions are equivalent:

      i. (M, ∇) is an affine Osserman space.

      ii. $(T^*M, \tilde{\nabla}_{\nabla,\Phi})$ is an affine Osserman space.

      iii. $(T^*M, g_{\nabla,\Phi})$ is a pseudo-Riemannian Osserman space for any Φ.

      iv. $(T^*M, g_{\nabla,\Phi})$ is a pseudo-Riemannian Osserman space for some Φ.

   (b) The following Assertions are equivalent:

      i. (M, ∇) is an affine Ivanov–Petrova space.

      ii. $(T^*M, \tilde{\nabla}_{\nabla,\Phi})$ is an affine Ivanova-Petrova space.

      iii. $(T^*M, g_{\nabla,\Phi})$ is a pseudo-Riemannian Ivanov–Petrova space for any Φ.

      iv. $(T^*M, g_{\nabla,\Phi})$ is a pseudo-Riemannian Ivanov–Petrova space for some Φ.

Proof. If the metric is self-dual, then we can use Equation (1.8.c) to express:

$$g_{11}(x^1, x^2, x_{1'}, x_{2'}) = x_{1'}^3 \mathcal{A} + x_{1'}^2 \mathcal{B} + x_{1'}^2 x_{2'} \mathcal{C} + x_{1'} x_{2'} \mathcal{D} + x_{1'} P + x_{2'} Q + \xi,$$

$$g_{22}(x^1, x^2, x_{1'}, x_{2'}) = x_{2'}^3 \mathcal{C} + x_{2'}^2 \mathcal{E} + x_{1'} x_{2'}^2 \mathcal{A} + x_{1'} x_{2'} \mathcal{F} + x_{1'} S + x_{2'} T + \eta,$$

$$g_{12}(x^1, x^2, x_{1'}, x_{2'}) = \tfrac{1}{2} x_{1'}^2 \mathcal{F} + \tfrac{1}{2} x_{2'}^2 \mathcal{D} + x_{1'}^2 x_{2'} \mathcal{A} + x_{1'} x_{2'}^2 \mathcal{C} + \tfrac{1}{2} x_{1'} x_{2'} (\mathcal{B} + \mathcal{E})$$
$$+ x_{1'} U + x_{2'} V + \gamma,$$

where the coefficients are smooth functions of (x^1, x^2). One checks that

$$\rho(\partial_{x_1}, \partial_{x_{1'}}) = \tfrac{1}{4}(16 x_{1'} \mathcal{A} + 8 x_{2'} \mathcal{C} + 5\mathcal{B} + \mathcal{E}), \qquad \rho(\partial_{x_1}, \partial_{x_{2'}}) = 2 x_{1'} \mathcal{C} + \mathcal{D},$$

$$\rho(\partial_{x_2}, \partial_{x_{2'}}) = \tfrac{1}{4}(8 x_{1'} \mathcal{A} + 16 x_{2'} \mathcal{C} + \mathcal{B} + 5\mathcal{E}), \qquad \rho(\partial_{x_2}, \partial_{x_{1'}}) = 2 x_{2'} \mathcal{A} + \mathcal{F},$$

and therefore, all the calligraphic letters vanish and thus, g is first order in $\{x_{1'}, x_{2'}\}$ and has the form of a deformed Riemannian extension. This establishes Assertion (1a).

We now prove Assertion (1b). For a general 4-dimensional pseudo-Riemannian manifold (M, g) with metric of neutral signature, it is easy to check that the characteristic polynomial $p_\lambda(\mathcal{R}(\pi))$ of $\mathcal{R}(\pi)$ is given by

$$p_\lambda(\mathcal{R}(\pi)) = \lambda^4 - \tfrac{1}{2} \operatorname{Tr}\{\mathcal{R}(\pi)^2\} \lambda^2 + \det\{\mathcal{R}(\pi)\}.$$

Thus, (M, g) is Ivanov–Petrova if and only if $\det\{\mathcal{R}(\pi)\}$ and $\mathrm{Tr}\{\mathcal{R}(\pi)^2\}$ do not depend on the oriented non-degenerate spacelike (respectively, mixed or timelike) 2-plane π (see Calviño-Louzao, García-Río, and Vázquez-Lorenzo [44]).

In the particular case of a Walker metric given in Remark 1.30, a straightforward calculation shows that the skew-symmetric curvature operator $\mathcal{R}(\pi)$ associated with any non-degenerate 2-plane π, when expressed with respect to the coordinate vectors $\{\partial_{x_i}, \partial_{x_{i'}}\}, i = 1, 2$, has the following matrix form:

$$\mathcal{R}(\pi) = \begin{pmatrix} F(\pi) & 0 \\ G(\pi) & -{}^tF(\pi) \end{pmatrix},$$

for suitably chosen 2×2 matrices $F(\pi)$ and $G(\pi)$. Thus:

$$\det\{\mathcal{R}(\pi)\} = \det\{F(\pi)\}^2 \text{ and } \mathrm{Tr}\{\mathcal{R}(\pi)^2\} = 2\,\mathrm{Tr}\{F(\pi)^2\}.$$

Therefore, the Walker metric of Remark 1.30 is Ivanov–Petrova if and only if $\det\{F(\pi)\}$ and $\mathrm{Tr}\{F(\pi)^2\}$ do not depend on the oriented non-degenerate spacelike (respectively, mixed or timelike) 2-plane π. Set

$$F(\pi) = \begin{pmatrix} f_{11}(\pi) & f_{12}(\pi) \\ f_{21}(\pi) & f_{22}(\pi) \end{pmatrix}.$$

Assume the Walker metric is Ivanov–Petrova and self-dual and hence, locally given by Equation (1.8.c). Let $\pi_1 = \mathrm{Span}\{\partial_{x_1}, \partial_{x_{1'}} + \lambda\partial_{x_{2'}}\}$ be a non-degenerate 2-plane. One computes:

$$f_{11}(\pi_1) = x_{1'}(\lambda\mathcal{C} + 3\mathcal{A}) + x_{2'}\mathcal{C} + \tfrac{1}{2}(\lambda\mathcal{D} + 2\mathcal{B}),$$
$$f_{12}(\pi_1) = x_{1'}\lambda\mathcal{A} + x_{2'}(\lambda\mathcal{C} + \mathcal{A}) + \tfrac{1}{4}(\lambda(\mathcal{B} + \mathcal{E}) + 2\mathcal{F}),$$
$$f_{21}(\pi_1) = x_{1'}\mathcal{C} + \tfrac{1}{2}\mathcal{D},$$
$$f_{22}(\pi_1) = x_{1'}(\lambda\mathcal{C} + \mathcal{A}) + x_{2'}\mathcal{C} + \tfrac{1}{4}(2\lambda\mathcal{D} + \mathcal{B} + \mathcal{E}).$$

Then $\partial_{x_{1'}}\partial_{x_{1'}}(\det\{F(\pi_1)\}) = 2\lambda^2\mathcal{C}^2 + 6\lambda\mathcal{A}\mathcal{C} + 6\mathcal{A}^2$, so $\mathcal{A} = \mathcal{C} = 0$ and one gets

$$\det\{F(\pi_1)\} = \tfrac{1}{4}\lambda^2\mathcal{D}^2 + \tfrac{1}{2}\lambda\mathcal{B}\mathcal{D} + \tfrac{1}{4}(\mathcal{B}^2 + \mathcal{B}\mathcal{E} - \mathcal{D}\mathcal{F}),$$

which implies that $\mathcal{D} = 0$. Hence Equation (1.8.c) simplifies in this instance to become:

$$g_{11}(x^1, x^2, x_{1'}, x_{2'}) = x_{1'}^2\mathcal{B} + x_{1'}P + x_{2'}Q + \xi,$$
$$g_{22}(x^1, x^2, x_{1'}, x_{2'}) = x_{2'}^2\mathcal{E} + x_{1'}x_{2'}\mathcal{F} + x_{1'}S + x_{2'}T + \eta,$$
$$g_{12}(x^1, x^2, x_{1'}, x_{2'}) = \tfrac{1}{2}x_{1'}^2\mathcal{F} + \tfrac{1}{2}x_{1'}x_{2'}(\mathcal{B} + \mathcal{E}) + x_{1'}U + x_{2'}V + \gamma.$$

Consider the non-degenerate 2-plane $\pi_2 = \mathrm{Span}\{\partial_{x_1} + \lambda\partial_{x_2}, \partial_{x_{1'}}\}$. We compute that:

$$f_{11}(\pi_2) = \tfrac{1}{2}(\lambda\mathcal{F} + 2\mathcal{B}), \quad f_{12}(\pi_2) = \tfrac{1}{2}\mathcal{F}$$
$$f_{21}(\pi_2) = \tfrac{1}{4}\lambda(\mathcal{B} + \mathcal{E}), \quad f_{22}(\pi_2) = \tfrac{1}{4}(2\lambda\mathcal{F} + \mathcal{B} + \mathcal{E}).$$

This implies that $\det\{F(\pi_2)\} = \tfrac{1}{4}\lambda^2\mathcal{F}^2 + \tfrac{1}{2}\lambda\mathcal{B}\mathcal{F} + \tfrac{1}{4}\mathcal{B}(\mathcal{B} + \mathcal{E})$. Consequently, $\mathcal{F} = 0$ and we may express g in the form:

$$g_{11}(x^1, x^2, x_{1'}, x_{2'}) = x_{1'}^2 \mathcal{B} + x_{1'} P + x_{2'} Q + \xi,$$
$$g_{22}(x^1, x^2, x_{1'}, x_{2'}) = x_{2'}^2 \mathcal{E} + x_{1'} S + x_{2'} T + \eta,$$
$$g_{12}(x^1, x^2, x_{1'}, x_{2'}) = \tfrac{1}{2} x_{1'} x_{2'} (\mathcal{B} + \mathcal{E}) + x_{1'} U + x_{2'} V + \gamma.$$

This permits us to compute:

$$\det\{F(\pi_2)\} = \tfrac{1}{4} \mathcal{B}(\mathcal{B} + \mathcal{E}), \qquad \mathrm{Tr}\{F(\pi_2)^2\} = \mathcal{B}^2 + \tfrac{1}{16}(\mathcal{B} + \mathcal{E})^2.$$

We also consider the non-degenerate 2-plane:

$$\pi_3 = \mathrm{Span}\{\partial_{x^{1'}} - \partial_{x^{2'}}, \partial_{x^{2'}} - \partial_{x_1} + \partial_{x_2}\} \text{ with } f_{11}(\pi_3) = \tfrac{1}{8}(5\mathcal{B} + \mathcal{E}),$$
$$f_{12}(\pi_3) = f_{21}(\pi_3) = -\tfrac{1}{8}(\mathcal{B} + \mathcal{E}), \ f_{22}(\pi_3) = \tfrac{1}{8}(\mathcal{B} + 5\mathcal{E}),$$
$$\det\{F(\pi_3)\} = \tfrac{1}{16}(\mathcal{B}^2 + \mathcal{E}^2 + 6\mathcal{B}\mathcal{E}), \text{ and } \mathrm{Tr}\{F(\pi_3)^2\} = \tfrac{1}{16}(7\mathcal{B}^2 + 7\mathcal{E}^2 + 6\mathcal{B}\mathcal{E}).$$

Comparing the values of $\det\{F(\pi_2)\}$ with $\det\{F(\pi_3)\}$ and of $\mathrm{Tr}\{F(\pi_2)^2\}$ with $\mathrm{Tr}\{F(\pi_3)^2\}$, we conclude that $\mathcal{E} = \mathcal{B}$. Hence, $\det\{F(\pi_3)\} = \tfrac{1}{2}\mathcal{B}^2$, which implies $\mathcal{B} = \kappa$ and we may express g in the form:

$$g_{11}(x^1, x^2, x_{1'}, x_{2'}) = x_{1'}^2 \kappa + x_{1'} P + x_{2'} Q + \xi,$$
$$g_{22}(x^1, x^2, x_{1'}, x_{2'}) = x_{2'}^2 \kappa + x_{1'} S + x_{2'} T + \eta,$$
$$g_{12}(x^1, x^2, x_{1'}, x_{2'}) = x_{1'} x_{2'} \kappa + x_{1'} U + x_{2'} V + \gamma.$$

To show that $\kappa = 0$, we consider the non-degenerate 2-plane:

$$\pi_4 = \mathrm{Span}\{\partial_{x_2} - c\partial_{x^{1'}} - \tfrac{1+b}{2}\partial_{x^{2'}}, \partial_{x_1} + \tfrac{1-a}{2}\partial_{x^{1'}}\}.$$

A straightforward computation then yields:

$$f_{11}(\pi_4) = \tfrac{1}{4}\left(x_{1'} x_{2'} 3\kappa^2 + (x_{1'} U + x_{2'} V)3\kappa + QS - UV + 4\kappa\gamma - 2P_2 + 2U_1\right),$$
$$f_{12}(\pi_4) = \tfrac{1}{4}\left(\kappa + \kappa\eta + S(V - P) - TU + U^2 + 2S_1 - 2U_2\right),$$
$$f_{21}(\pi_4) = \tfrac{1}{4}\left(\kappa - \kappa\xi + Q(T - U) + PV - V^2 - 2Q_2 + 2V_1\right),$$
$$f_{22}(\pi_4) = \tfrac{1}{4}\left(x_{1'} x_{2'} 3\kappa^2 + (x_{1'} U + x_{2'} V)3\kappa - QS + UV + 2\kappa\gamma + 2T_1 - 2V_2\right).$$

We now verify that $\partial_{x^{1'}} \partial_{x^{1'}} \partial_{x^{2'}} \partial_{x^{2'}} (\det\{F(\pi_4)\}) = \tfrac{9}{4}\kappa^4$. This shows that $\kappa = 0$, and hence, g has the form of a deformed Riemannian extension as desired; Assertion (1b) follows.

We now prove Assertion (2). The equivalence of Assertions (2-a-i) and (2-a-ii) and of Assertions (2-b-i) and (2-b-ii) now follows from Lemma 2.14. Also, clearly Assertion (2-a-ii) implies Assertion (2-a-iii) and, similarly, Assertion (2-b-ii) implies Assertion (2-b-iii).

It is immediate that Assertion (2-a-iii) implies Assertion (2-a-iv) and that Assertion (2-b-iii) implies Assertion (2-b-iv). Suppose Assertion (2-a-iv) holds so that there exists Φ for which $(T^*M, g_{\nabla, \Phi})$ is pseudo-Riemannian Osserman. Let $S := \mathrm{Spec}\{\tilde{\mathcal{J}}(\tilde{X})\}$ be the common spectrum for any $\tilde{X} \in S^+(T_{\tilde{P}} T^*M, g_{\nabla, \Phi})$. For $t > 0$, set $\tilde{X}_t := t^{-1}\partial_{x_1} + t\partial_{x^{1'}}$. Then:

$$g_{\nabla, \Phi}(\tilde{X}_t, \tilde{X}_t) = 1 + O(t^{-1}) \text{ so}$$
$$S = (1 + O(t^{-1})) \cdot \mathrm{Spec}\{\tilde{\mathcal{J}}(\tilde{X}_t)\} = (1 + O(t^{-1}))t^{-2} \mathrm{Spec}\{\mathcal{J}(\partial_{x_1})\}.$$

Taking the limit as $t \to \infty$ yields $S = \{0\}$ and hence, $(M, \tilde{\nabla}_{\nabla, \Phi})$ is affine Osserman. Similarly suppose Assertion (2-b-iv) holds so that there exists Φ for which $(T^*M, g_{\nabla, \Phi})$ is pseudo-Riemannian Ivanov–Petrova and let $S := \text{Spec}\{\tilde{\mathcal{R}}(\tilde{X}, \tilde{Y})\}$ be the common spectrum for any $\{\tilde{X}, \tilde{Y}\}$, an orthonormal basis for a spacelike 2-plane. Let $X = \partial_{x_1}$ and $Y = \partial_{x_2}$. Set

$$\tilde{X}_t := t^{-1} \partial_{x_1} + t \partial_{x_{1'}} \text{ and } \tilde{Y}_t := t^{-1} \partial_{x_2} + t \partial_{x_{2'}}.$$

Then

$$g_{\nabla, \Phi}(\tilde{X}_t, \tilde{X}_t) g_{\nabla, \Phi}(\tilde{Y}_t, \tilde{Y}_t) = g_{\nabla, \Phi}(\tilde{X}_t, \tilde{Y}_t)^2 = 1 + O(t^{-1}) \text{ so}$$
$$S = (1 + O(t^{-1})) \cdot \text{Spec}\{\tilde{\mathcal{R}}(\tilde{X}_t, \tilde{Y}_t)\} = (1 + O(t^{-1})) t^{-2} \text{Spec}\{\mathcal{R}(\partial_{x_1}, \partial_{x_2})\}.$$

Taking the limit as $t \to \infty$ yields $S = \{0\}$, so $(M, \tilde{\nabla}_{\nabla, \Phi})$ is affine Ivanov–Petrova. $\square$

In fact, we have proved a bit more. The following is a scholium to the proof and is a local version of Theorem 2.15:

Theorem 2.16   *Let (M, ∇) be an affine manifold. Let $\tilde{\nabla}_{\nabla, \Phi}$ be the Levi-Civita connection of $(T^*M, g_{\nabla, \Phi})$. Let $P \in M$ and let $\omega \in T_P^*M$.*

1. *The following Assertions are equivalent:*

   (a) *(M, ∇) is an affine Osserman space at P.*

   (b) *$(T^*M, \tilde{\nabla}_{\nabla, \Phi})$ is an affine Osserman space at (P, ω) for any ω and any Φ.*

   (c) *$(T^*M, g_{\nabla, \Phi})$ is a pseudo-Riemannian Osserman space at (P, ω) for any ω and any Φ.*

   (d) *$(T^*M, g_{\nabla, \Phi})$ is a pseudo-Riemannian Osserman space at (P, ω) for some (ω, Φ).*

2. *The following Assertions are equivalent:*

   (a) *(M, ∇) is an affine Ivanov–Petrova space at P.*

   (b) *$(T^*M, \tilde{\nabla}_{\nabla, \Phi})$ is an affine Ivanova-Petrova space at (P, ω) for any ω and for any Φ.*

   (c) *$(T^*M, g_{\nabla, \Phi})$ is a pseudo-Riemannian Ivanov–Petrova space at (P, ω) for any ω and for any Φ.*

   (d) *$(T^*M, g_{\nabla, \Phi})$ is a pseudo-Riemannian Ivanov–Petrova space at (P, ω) for some ω and for some Φ.*

Note that self-dual Walker metrics that have nilpotent Ricci operator are not necessarily deformed Riemannian extensions. For instance, the self-dual Walker metric given by Remark 1.30 where

$$g_{11} = 0, \qquad g_{22} = x_{1'} x_{2'} \mathcal{A}(x^1, x^2), \qquad g_{12} = \tfrac{1}{2}(x_{1'})^2 \mathcal{A}(x^1, x^2),$$

has a 2-step nilpotent Ricci operator, but does not correspond to any deformed Riemannian extension.

We may apply Theorem 2.6 (4) to the situation in Theorem 2.15 (1). Any Ricci flat self-dual Walker manifold (M, g) is locally isometric to the cotangent bundle $T^*\Sigma$ of an affine surface (Σ, ∇) equipped with the deformed Riemannian extension $g_{\nabla, \Phi}$. The affine connections in question are locally parametrized by a function $\theta(x^1, x^2)$; there is no restriction on the symmetric $(0, 2)$-tensor field Φ.

While any 4-dimensional Riemannian Ivanov–Petrova manifold is necessarily locally conformally flat, there are examples in neutral signature of 4-dimensional Ivanov–Petrova manifolds that are neither self-dual nor anti-self-dual, as shown in Calviño-Louzao, García-Río, and Vázquez-Lorenzo [44]. For instance, any Walker metric of the form:

$$g_{11}(x^1, x^2, x_{1'}, x_{2'}) = x_{2'}^2 P(x^1) + x_{1'} S(x^1) + x_{2'} T(x^1, x^2) + \xi(x^1, x^2),$$
$$g_{22}(x^1, x^2, x_{1'}, x_{2'}) = x_{2'} \kappa + \eta(x^1, x^2), \quad g_{12}(x^1, x^2, x_{1'}, x_{2'}) = 0$$

is Ivanov–Petrova, but neither self-dual nor anti-self-dual. If the Walker metric is self-dual, then it must be a deformed Riemannian extension (see, for example, Calviño-Louzao, García-Río, and Vázquez-Lorenzo [45]).

CHAPTER 3

The Geometry of Modified Riemannian Extensions

Let Φ be a symmetric $(0, 2)$-tensor field on an affine manifold (M, ∇) of dimension $\bar{m}$ and let σ be the natural projection from T^*M to M. We set $T = S = \mathrm{Id}$ and adopt the notation of Section 1.7. The modified Riemannian extension $g_{\nabla, \Phi, c}$ is the metric of neutral signature $(\bar{m}, \bar{m})$ on the cotangent bundle given in a coordinate free fashion by taking:

$$g_{\nabla, \Phi, c} = c\, \iota\, \mathrm{Id} \circ \iota\, \mathrm{Id} + g_\nabla + \sigma^*\Phi.$$

In local coordinates $(\vec{x}, \vec{x}')$, this takes the form:

$$g_{\nabla, \Phi, c} = 2\, dx^i \circ dx_{i'} + \left(c\, x_{i'} x_{j'} + \Phi_{ij}(\vec{x}) - 2x_{k'} \Gamma_{ij}{}^k(\vec{x}) \right) dx^i \circ dx^j.$$

Modified Riemannian extensions are useful in describing 4-dimensional Osserman metrics, as we shall see in Section 3.1. In Section 3.2, we will explore the setting where the underlying affine connection is flat since this provides local models for the para-complex space forms. This motivates the discussion in Section 3.3 where we use non-flat affine manifolds to construct higher-dimensional Osserman metrics on the cotangent bundle, which are not nilpotent and which have non-trivial Jordan normal form. These examples will be explored in greater detail in Section 3.4. We conclude Chapter 3 in Section 3.5 with a discussion of (semi) para-complex Osserman manifolds.

3.1 FOUR-DIMENSIONAL OSSERMAN MANIFOLDS AND MODELS

The work of Blažić, Bokan, and Gilkey [16] and of García-Río, Kupeli, and Vázquez-Abal [88] shows that any Lorentzian Osserman manifold has constant sectional curvature. Furthermore, any Osserman metric is Einstein. Since 3-dimensional Einstein metrics have constant sectional curvature, any Osserman manifold of dimension three has constant sectional curvature. A result of Chi [53] (see Theorem 3.3 below) shows that any Osserman manifold of signature $(0, 4)$ is locally isometric to a real space form or to a complex space form. Thus, the first non-trivial case arises in signature $(2, 2)$ and such manifolds are not completely classified, although there are many examples and some results in this direction.

There is a special relationship between self-duality and the Osserman condition in signature $(0, 4)$ and $(2, 2)$ that will play an important role in our discussion. Let W be the Weyl conformal curvature tensor defined in Equation (1.5.f):

$$W(X, Y, Z, T) := R(X, Y, Z, T) + \frac{1}{(m-1)(m-2)}\tau\{g(Y, Z)g(X, T) - g(X, Z)g(Y, T)\}$$

$$-\frac{1}{m-2}\{g(Y, Z)\rho(X, T) - g(X, Z)\rho(Y, T) + g(X, T)\rho(Y, Z) - g(Y, T)\rho(X, Z)\}.$$

Let W be the associated conformal curvature operator. Fix an orientation. If $m = 4$, then we can decompose $W = W^+ + W^-$ into the *self-dual* and *anti-self-dual* Weyl conformal curvature operators. The manifold (M, g) is said to be *self-dual* if $W^- = 0$; this is a conformally invariant property. If $W^+ = 0$, we will usually reverse the orientation to simplify the discussion. We refer to Gilkey, Swann, and Vanhecke [114] in signature $(0, 4)$ and to Brozos-Vázquez, García-Río, and Vázquez-Lorenzo [33] in signature $(2, 2)$ for the proof of the following result:

Lemma 3.1   *Let $\mathfrak{M} = (V, \langle \cdot, \cdot \rangle, A)$ be a curvature model of signature $(2, 2)$ or $(0, 4)$. Then $\mathfrak{M}$ is Osserman if and only if $\mathfrak{M}$ is Einstein and self-dual.*

OSSERMAN MANIFOLDS OF SIGNATURE $(0, 4)$

The study of Osserman metrics is, in general, a 2-step process. First one works in the purely algebraic setting and classifies the set of all Riemannian Osserman curvature models of a given signature. One then passes to the geometric setting to determine which of the Riemannian Osserman curvature models can be geometrically realized by an Osserman manifold. Although we are primarily interested in signature $(2, 2)$ geometry, it is worth illustrating this process first in the Riemannian setting. Let $\mathfrak{M} = (V, \langle \cdot, \cdot \rangle, A)$ be an Osserman curvature model of signature $(0, 4)$. The curvature tensor of constant sectional curvature is given by:

$$\mathcal{A}_0(x, y)z := \langle y, z \rangle x - \langle x, z \rangle y.$$

More generally, if Ψ is a skew-symmetric linear operator on $(V, \langle \cdot, \cdot \rangle)$, then we may define an algebraic curvature tensor by setting:

$$\mathcal{A}_\Psi(x, y)z := \langle \Psi y, z \rangle \Psi x - \langle \Psi x, z \rangle \Psi y - 2\langle \Psi x, y \rangle \Psi z.$$

We say that a collection $\{\Psi_1, \Psi_2, \Psi_3\}$ of skew-adjoint endomorphisms of V gives $(V, \langle \cdot, \cdot \rangle)$ a *quaternion structure* if we have

$$\Psi_i \Psi_j + \Psi_j \Psi_i = -2\delta_i^j \text{ Id and } \Psi_3 = \Psi_2 \Psi_1.$$

Theorem 1.52 specializes to the 4-dimensional setting to give:

Theorem 3.2   *Let $(V, \langle \cdot, \cdot \rangle, A)$ be a 4-dimensional Riemannian Osserman curvature model. Then there exists a quaternion structure $\{\Psi_1, \Psi_2, \Psi_3\}$ on $(V, \langle \cdot, \cdot \rangle)$ and there exist constants $\{\alpha_0, \alpha_1, \alpha_2, \alpha_3\}$ so that $\mathcal{A} = \alpha_0 \mathcal{A}_0 + \alpha_1 \mathcal{A}_{\Psi_1} + \alpha_2 \mathcal{A}_{\Psi_2} + \alpha_3 \mathcal{A}_{\Psi_3}.$*

We now center our attention in the geometric setting to prove a result of Chi [53] classifying Riemannian Osserman manifolds in dimension four. Lemma 3.1 plays a crucial role in our analysis.

Theorem 3.3 *If (M, g) is a 4-dimensional Riemannian Osserman manifold, then (M, g) is locally isometric either to a real space form or to a complex space form.*

Proof. Let $\mathfrak{M}$ be an Osserman curvature model of signature $(0, 4)$; $\mathfrak{M}$ has the form given in Theorem 3.2. By Lemma 3.1, $\mathfrak{M}$ is self-dual and Einstein. Consequently, all the information we shall need is encoded by the self-dual Weyl curvature operator $\mathcal{W}^+$. If $\mathcal{W}^+$ vanishes identically, then $\mathfrak{M}$ is conformally flat and $\mathfrak{M}$ has constant sectional curvature, so $\mathfrak{M}$ is geometrically represented by a space form. We therefore assume $\mathcal{W}^+$ non-zero. Since $\mathcal{W}^+$ is traceless, $\mathcal{W}^+$ has either two or three distinct eigenvalues.

Suppose there exists an Osserman manifold (M, g) of signature $(0, 4)$ that is not conformally flat and which realizes $\mathfrak{M}$. The eigenvalues of $\mathcal{W}^+$ are in correspondence with the eigenvalues of the Jacobi operator and hence, are constant on (M, g). Thus, since $\mathcal{W}^+$ is self-adjoint, $\langle \mathcal{W}^+, \mathcal{W}^+ \rangle$ is constant. Suppose that $\mathcal{W}^+$ has exactly two distinct eigenvalues. By the generalized Goldberg–Sachs theorem (see, for example, Apostolov [8] and Ivanov and Zamkovoy [135]), there exists a distinguished 2-form Ω on M with $\langle \Omega, \Omega \rangle = 2$ corresponding to the eigenvalue of $\mathcal{W}^+$ which has multiplicity 1. Furthermore, Ω is the Kähler form of an almost Hermitian structure (g, J_-); the work of Derdzinski [64, Proposition 5] shows that (g, J_-) is actually locally conformally Kähler with conformal factor $(24\langle \mathcal{W}^+, \mathcal{W}^+ \rangle)^{1/3}$. Since $\langle \mathcal{W}^+, \mathcal{W}^+ \rangle$ is constant, (g, J_-) is in fact Kähler. Bryant [39] has shown that 4-dimensional Bochner flat Kähler metrics are characterized by the fact that the anti-self-dual Weyl curvature operator $\mathcal{W}^-$ vanishes. The Einstein condition now shows that (g, J_-) is a Kähler structure of constant holomorphic sectional curvature and thus, as desired, (M, g) is a complex space form.

Finally, suppose that $\mathcal{W}^+$ has 3-distinct eigenvalues. We shall use results established by Derdzinski [64] to obtain a contradiction. Let Ω_i be the three orthogonal eigenvectors of $\mathcal{W}^+$ corresponding to the three different eigenvalues μ_i. Let $\star$ be the Hodge operator. Since $\nabla \star = 0$, the decomposition $\Lambda^2(M) = \Lambda^+(M) \oplus \Lambda^-(M)$ is parallel. Consequently, there exist 1-forms a, b, c such that

$$
\begin{aligned}
\nabla \Omega_1 &= & c \otimes \Omega_2 - b \otimes \Omega_3, \\
\nabla \Omega_2 &= -c \otimes \Omega_1 & + a \otimes \Omega_3, \\
\nabla \Omega_3 &= b \otimes \Omega_1 - a \otimes \Omega_2.
\end{aligned}
$$

We use the second Bianchi identity to see that $\delta \mathcal{W}^+ = 0$, or, equivalently,

$$
\begin{aligned}
d\mu_1 &= (\mu_1 - \mu_2)\Omega_3 c + (\mu_1 - \mu_3)\Omega_2 b, \\
d\mu_2 &= (\mu_2 - \mu_1)\Omega_3 c + (\mu_2 - \mu_3)\Omega_1 a, \\
d\mu_3 &= (\mu_3 - \mu_1)\Omega_2 b + (\mu_3 - \mu_2)\Omega_1 a.
\end{aligned}
$$

Recall that a *hyper Kähler structure* on a Riemannian manifold (M, g) is a triple of parallel orthogonal almost complex structures J_1, J_2, J_3, such that $J_1 J_2 = -J_2 J_1 = J_3$; therefore, (g, J_i) is Kähler for all $i = 1, 2, 3$. In particular, any hyper Kähler manifold is Ricci flat (see Hitchin [127] for further details). In the setting at hand, we may now conclude either that at least two of the eigenvalues μ_i are equal (which is a contradiction) or, alternatively, that the forms Ω_i are parallel for all i. In this instance, $(\Omega_1, \Omega_2, \Omega_3)$ defines a hyper Kähler structure on M, which is a contradiction.   □

OSSERMAN MODELS OF SIGNATURE (2, 2)

We say that $\{\Psi_1, \Psi_2, \Psi_3\}$ is a *Clifford module structure* of type $(1, 1)$ on an inner product space $(V, \langle \cdot, \cdot \rangle)$ of signature $(2, 2)$ if the Ψ_i are skew-adjoint operators satisfying:

$$\Psi_3 = \Psi_2 \Psi_1, \quad \Psi_1^2 = -\,\mathrm{Id}, \quad \Psi_2^2 = \Psi_3^2 = \mathrm{Id}, \text{ and } \Psi_i \Psi_j + \Psi_j \Psi_i = 0 \text{ if } i \neq j.$$

We refer to Karoubi [138] for further details concerning Clifford modules. An interesting aspect of Clifford module structures of type $(1, 1)$ is that they give rise to skew-symmetric 2-step nilpotent operators $\Psi_1 - \Psi_j$ for $j = 2, 3$ and to a skew-symmetric almost product structure (after rescaling) $\Psi_2 - \Psi_3$. Theorem 3.2 generalizes to this setting; a complete description of all Osserman algebraic curvature tensors in the neutral signature is given by the following result of Blažić *et al.* [21]:

Theorem 3.4   *A curvature model* $(V, \langle \cdot, \cdot \rangle, A)$ *of signature* $(2, 2)$ *is Osserman if and only if there exists a Clifford module structure* $\{\Psi_1, \Psi_2, \Psi_3\}$ *of type* $(1, 1)$ *on* V, *and there exist constants* α_i *and* β_{ij} *so that*

$$A = \alpha_0 A_0 + \sum \alpha_i A_{\Psi_i} + \sum_{i<j} \beta_{ij} [A_{\Psi_i} + A_{\Psi_j} - A_{(\Psi_i - \Psi_j)}].$$

Let $\mathfrak{M} = (V, \langle \cdot, \cdot \rangle, A)$ be an Osserman curvature model of signature $(2, 2)$. If X is not a null vector, then the induced metric on $X^\perp$ has Lorentzian signature and thus, $\mathcal{J}(X)$ can have complex eigenvalues. Any 3×3 matrix is either diagonalizable (this gives rise to Type Ia below), or has complex roots (this gives rise to Type Ib below), or has a 2×2 Jordan block with real eigenvalues (this gives rise to Type II below), or has a 3×3 Jordan block with a real eigenvalue (this gives rise to Type III below). Consequently, in this setting, the Jacobi operator $\mathcal{J}(X) = \mathcal{R}(\cdot, X)X$, viewed as an endomorphism of $X^\perp$, corresponds to one of the following possibilities:

$$\begin{pmatrix} \alpha & 0 & 0 \\ 0 & \beta & 0 \\ 0 & 0 & \gamma \end{pmatrix}, \quad \begin{pmatrix} \alpha & -\beta & 0 \\ \beta & \alpha & 0 \\ 0 & 0 & \gamma \end{pmatrix}, \quad \begin{pmatrix} \alpha & 0 & 0 \\ 0 & \beta & 0 \\ 0 & 1 & \beta \end{pmatrix}, \quad \begin{pmatrix} \alpha & 0 & 0 \\ 1 & \alpha & 0 \\ 0 & 1 & \alpha \end{pmatrix}. \qquad (3.1.a)$$

$$\textit{Type Ia} \qquad\qquad \textit{Type Ib} \qquad\qquad \textit{Type II} \qquad\qquad \textit{Type III}$$

The possible Jordan normal forms of the Jacobi operators in Equation (3.1.a) are in a one-to-one correspondence with the possible Jordan normal forms of the corresponding self-dual Weyl curvature operator $\mathcal{W}^+$. There is a complete algebraic classification due to Blažić, Bokan, and Rakić [17]—see also the discussion in García-Río *et al.* [86]. Let $\mathfrak{M}$ be an Osserman curvature model of signature $(2, 2)$. In what follows, we only give the non-zero components of A and $\mathcal{W}^+$. We shall consider an orthonormal basis $\{e_1, e_2, e_3, e_4\}$, where $\{e_1, e_2\}$ are timelike and $\{e_3, e_4\}$ are spacelike. The orthonormal basis for Λ^+ given in Equation (1.8.a) has signature (2, 1):

$$E_1^+ = \frac{e^1 \wedge e^2 + e^3 \wedge e^4}{\sqrt{2}}, \quad E_2^+ = \frac{e^1 \wedge e^3 + e^2 \wedge e^4}{\sqrt{2}}, \quad E_3^+ = \frac{e^1 \wedge e^4 - e^2 \wedge e^3}{\sqrt{2}},$$

$$\langle E_1^+, E_1^+ \rangle = 1, \qquad \langle E_2^+, E_2^+ \rangle = -1, \qquad \langle E_3^+, E_3^+ \rangle = -1.$$

Case Ia: The Jacobi operators are diagonalizable. There is an orthonormal basis for V where $\{e_1, e_2\}$ are timelike and $\{e_3, e_4\}$ are spacelike so:

$$A_{1221} = A_{4334} = \alpha, \qquad A_{1331} = A_{4224} = -\beta, \qquad A_{1441} = A_{3223} = -\gamma,$$

$$A_{1234} = (-2\alpha + \beta + \gamma)/3, \quad A_{1423} = (\alpha + \beta - 2\gamma)/3, \quad A_{1342} = (\alpha - 2\beta + \gamma)/3,$$

$$\mathcal{W}^+ = \frac{2}{3} \begin{pmatrix} 2\alpha - \beta - \gamma & 0 & 0 \\ 0 & -\alpha + 2\beta - \gamma & 0 \\ 0 & 0 & -\alpha - \beta + 2\gamma \end{pmatrix}.$$

Case Ib: The Jacobi operators have a complex eigenvalue $\alpha \pm \sqrt{-1}\beta$. There is an orthonormal basis for V where $\{e_1, e_2\}$ are timelike and $\{e_3, e_4\}$ are spacelike so:

$$A_{1221} = A_{4334} = \alpha, \qquad A_{1331} = A_{4224} = -\alpha, \quad A_{1441} = A_{3223} = -\gamma,$$

$$A_{2113} = A_{2443} = -\beta, \quad A_{1224} = A_{1334} = \beta, \qquad A_{1234} = (-\alpha + \gamma)/3,$$

$$A_{1423} = 2(\alpha - \gamma)/3, \qquad A_{1342} = (-\alpha + \gamma)/3,$$

$$\mathcal{W}^+ = \begin{pmatrix} -\frac{2}{3}(\alpha - \gamma) & -2\beta & 0 \\ 2\beta & -\frac{2}{3}(\alpha - \gamma) & 0 \\ 0 & 0 & \frac{4}{3}(\alpha - \gamma) \end{pmatrix}.$$

Case II: The minimal polynomial of the Jacobi operators has a double root. There is an orthonormal basis for V where $\{e_1, e_2\}$ are timelike and $\{e_3, e_4\}$ are spacelike so:

$$A_{1221} = A_{4334} = \varepsilon\left(\alpha - \frac{1}{2}\right), \qquad A_{2113} = A_{2443} = -\frac{\varepsilon}{2}, \qquad A_{1441} = A_{3223} = -\beta,$$

$$A_{1331} = A_{4224} = -\varepsilon\left(\alpha + \frac{1}{2}\right), \qquad A_{1224} = A_{1334} = \frac{\varepsilon}{2}, \qquad A_{1423} = \frac{2}{3}(\varepsilon\alpha - \beta),$$

$$A_{1234} = \frac{1}{3}\left(\varepsilon\left(-\alpha + \frac{3}{2}\right) + \beta\right), \qquad A_{1342} = \frac{1}{3}\left(\varepsilon\left(-\alpha - \frac{3}{2}\right) + \beta\right),$$

$$\mathcal{W}^+ = \begin{pmatrix} -\frac{2(\alpha-\beta)}{3} + 1 & -1 & 0 \\ 1 & -\frac{2(\alpha-\beta)}{3} - 1 & 0 \\ 0 & 0 & \frac{4(\alpha-\beta)}{3} \end{pmatrix} \text{ where } \varepsilon = \pm 1.$$

Case III: The minimal polynomial of the Jacobi operators has a triple root. There is an orthonormal basis for V where $\{e_1, e_2\}$ are timelike and $\{e_3, e_4\}$ are spacelike so:

$$A_{1221} = A_{4334} = \alpha, \qquad A_{1331} = A_{4224} = -\alpha, \qquad A_{1441} = A_{3223} = -\alpha,$$

$$A_{2114} = A_{2334} = -\tfrac{\sqrt{2}}{2}, \qquad A_{3114} = -A_{3224} = \tfrac{\sqrt{2}}{2}, \qquad A_{1223} = A_{1443} = \tfrac{\sqrt{2}}{2},$$

$$A_{1332} = -A_{1442} = \tfrac{\sqrt{2}}{2}, \quad \mathcal{W}^+ = \begin{pmatrix} 0 & 0 & \sqrt{2} \\ 0 & 0 & \sqrt{2} \\ -\sqrt{2} & \sqrt{2} & 0 \end{pmatrix}.$$

A triple of skew-adjoint operators $\{\Psi_1, \Psi_2, \Psi_3\}$ forms a *para-quaternionic structure* if

$$\Psi_1^2 = -\operatorname{id}, \quad \Psi_2^2 = \operatorname{id}, \quad \Psi_3^2 = \operatorname{id}, \quad \Psi_i \Psi_j + \Psi_j \Psi_i = 0 \text{ for } i \neq j.$$

The following result of García-Río *et al.* [86] arises from the classification given above:

Theorem 3.5   *Let $\mathfrak{M} = (V, \langle \cdot, \cdot \rangle, A)$ be a curvature model of signature $(2, 2)$.*

1. *The following conditions are equivalent:*

   (a) *$\mathfrak{M}$ is spacelike Osserman or timelike Osserman or null Osserman.*

   (b) *$\mathfrak{M}$ is Einstein and self-dual.*

   (c) *$\mathfrak{M}$ is spacelike Jordan–Osserman or timelike Jordan–Osserman.*

2. *$\mathfrak{M}$ is null Jordan–Osserman if and only if $\mathfrak{M}$ is of Type Ia and if one of the following conditions holds for some constants $\kappa_i \in \mathbb{R}$:*

   (a) *A has constant sectional curvature, i.e., $A = \kappa_0 A_0$.*

   (b) *There exists a Hermitian structure J_- so $A = \kappa_0 A_0 + \kappa_1 A_{J_-}$ for $\kappa_1 \neq 0$.*

   (c) *There exists a para-Hermitian structure J_+ so $A = \kappa_1 A_{J_+}$ for $\kappa_1 \neq 0$.*

   (d) *There exists a para-quaternionic structure $\{\Psi_1, \Psi_2, \Psi_3\}$ so that*

   $$A = \kappa_1 A_{\Psi_1} + \kappa_2 A_{\Psi_2} + \kappa_3 A_{\Psi_3}$$

   where $\kappa_2 \kappa_3 (\kappa_2 + \kappa_1)(\kappa_3 + \kappa_1) > 0$ and where $\{3\kappa_1, -3\kappa_2, -3\kappa_3\}$ are all distinct.

OSSERMAN MANIFOLDS OF SIGNATURE $(2, 2)$

We refer to García-Río and Vázquez-Lorenzo [92] for the proof of the following result:

Theorem 3.6   *Any locally symmetric 4-dimensional Osserman manifold is either a real, a complex, or a para-complex space form or is locally isometric to the following example, which has a 2-step nilpotent Jacobi operator:*

$$g = 2dx^1 \circ dx^2 + 2dx^3 \circ dx^4 \pm (x^1)^2 dx^4 \otimes dx^4.$$

We now examine seriatim the algebraic cases in Equation (3.1.a).

Type Ia: Diagonalizable Jacobi Operator

We have the following result of Blažić, Bokan, and Rakić [17]:

Theorem 3.7   *Any Type Ia Osserman manifold is locally isometric to a real, a complex, or a para-complex space form.*

Proof. Let (M, g) be a Type Ia Osserman manifold of signature $(2, 2)$. If $\mathcal{W}^+$ has only one eigenvalue and is diagonalizable, then the metric has constant sectional curvature. If $\mathcal{W}^+$ has exactly two distinct eigenvalues, let Ω be an eigenvector corresponding to the eigenvalue of multiplicity 1. Since the induced metric on Λ^+ has Lorentzian signature, one may normalize the choice of Ω by requiring that $\langle \Omega, \Omega \rangle = \pm 2$ (Ω is not null). If Ω is spacelike, then Ω defines an indefinite almost Hermitian structure (g, J_-), while, if Ω is timelike, then Ω defines an almost para-Hermitian structure (g, J_+). A generalization of the Goldberg–Sachs theorem due to Apostolov [8] and to Ivanov and Zamkovoy [135] now shows that such structures are locally conformally (para)-Kähler. As in the Riemannian case, which was discussed above, the conformal factor is given by the norm of the self-dual Weyl curvature operator and hence, it is constant as (M, g) is assumed Osserman. Therefore, the structures associated to Ω are Kähler (with respect to the opposite orientation) or para-Kähler, depending on the causality of Ω. The vanishing of $\mathcal{W}^-$ implies that the Bochner curvature tensor vanishes so the manifold is Bochner flat (see, for example, Bryant [39]), and thus, the metric corresponds either to an indefinite complex space form or to a para-complex space form. Finally, proceeding as in our discussion of 4-dimensional Riemannian Osserman manifolds, one shows that the Jacobi operators may not have 3-distinct eigenvalues.   □

Type Ib: Complex Eigenvalues

Blažić, Bokan, and Rakić [17] also showed:

Theorem 3.8   *There are no Type Ib Osserman manifolds.*

Proof. Although Type Ib curvature models exist, they are not geometrically realizable as Type Ib Osserman manifolds. Suppose (M, g) is a Type Ib Osserman manifold of signature $(2, 2)$. We argue for a contradiction. Recall that a pseudo-Riemannian manifold is said to be *curvature homogeneous of order k* if, for each pair of points $P, Q \in M$, there exists an isometry Θ_{PQ} from $T_P M$ to $T_Q M$ so $\Theta_{PQ}^* \nabla^j R_Q = \nabla^j R_P$ for $0 \leq j \leq k$. As the Jordan normal form of the Jacobi operator may change from point to point, a 4-dimensional Osserman manifold is not necessarily curvature homogeneous; the Jordan normal form of the Jacobi operator is constant, of course, on $S^\pm(M, g)$ by Theorem 3.5. However, if (M, g) is Type Ib, then the Jordan normal form must be constant on all of M and (M, g) is 0-curvature homogeneous; a similar argument holds for the self-dual Weyl curvature operator.

Derdzinski [65] showed that, if (M, g) is a signature $(2, 2)$ Einstein manifold with a complex diagonalizable Weyl curvature operator, then M is either locally symmetric or is locally isometric to a Lie group with a left-invariant metric. We describe this example as follows. Let (x^1, x^2, x^3, x^4) be the usual coordinates on $\mathbb{R}^4$. Let

$$g = e^{k(x^1)^2} \cos k(x^1)^2 \sqrt{3}(dx^1 \circ dx^1 - dx^4 \circ dx^4) + dx^2 \circ dx^2 - e^{-2k(x^1)^2} dx^3 \circ dx^3$$
$$- 2e^{k(x^1)^2} \sin k(x^1)^2 \sqrt{3} dx^1 \circ dx^4.$$

The restriction of the curvature operator $\mathcal{R} : \Lambda^2 \to \Lambda^2$ to the spaces of (anti)-self-dual 2-forms defines maps $\mathcal{R}^\pm : \Lambda^\pm \to \Lambda^\pm$, which have constant eigenvalues $\{-k^2, \frac{1}{2}(k^2 \pm \sqrt{-3}k^2)\}$ for $k \neq 0$. Consequently, none of these metrics is self-dual, and hence, no 4-dimensional Osserman metric may have a Jacobi operator of Type Ib.   □

Type II and Type III Osserman Manifolds

Theorem 3.6 exhibits a locally symmetric Osserman manifold with a 2-step nilpotent Jacobi operator. We refer to the discussion in Bonome *et al.* [26], in García-Río, Kupeli, and Vázquez-Lorenzo [90], and in García-Río, Vázquez-Abal, and Vázquez-Lorenzo [91] for other examples of 2-step and 3-step nilpotent Osserman manifolds of signature (2, 2); the classification is far from complete.

However, if the Jacobi operators are non-nilpotent, then the work of Blažić, Bokan, and Rakić [17] provides a complete description in the Type II setting. Such a metric is necessarily of Walker type. A systematic analysis of self-duality conditions for Walker metrics was carried out by Davidov and Muskarov [62] and by Díaz-Ramos, García-Río, and Vázquez-Lorenzo [71], thus leading to the following result which motivates the construction of the new examples of Osserman metrics that we shall give in the next sections and which is closely related to Lemma 1.39:

Theorem 3.9   *Let (M, g) be a Type II Osserman manifold. Then one of the following two possibilities holds:*

1. *The Jacobi operators are 2-step nilpotent.*

2. *The scalar curvature τ is not zero. In this situation, we have:*

   (a) *There exist local coordinates $(x^1, x^2, x_{1'}, x_{2'})$ and there exist $\{P, Q, S, T, U, V\}$, which are functions of (x^1, x^2) so:*

   $$g = 2dx^1 \circ dx_{1'} + 2dx^2 \circ dx_{2'} + g_{11}dx^1 \circ dx^1 + g_{22}dx^2 \circ dx^2 + 2g_{12}dx^1 \circ dx^2,$$
   $$g_{11} = (x_{1'})^2 \tfrac{\tau}{6} + x_{1'}P + x_{2'}Q + \tfrac{6}{\tau}\{Q(T - U) + V(P - V) - 2(Q_2 - V_1)\},$$
   $$g_{22} = (x_{2'})^2 \tfrac{\tau}{6} + x_{1'}S + x_{2'}T + \tfrac{6}{\tau}\{S(P - V) + U(T - U) - 2(S_1 - U_2)\},$$
   $$g_{12} = x_{1'}x_{2'}\tfrac{\tau}{6} + x_{1'}U + x_{2'}V + \tfrac{6}{\tau}\{-QS + UV + T_1 - U_1 + P_2 - V_2\}.$$

   (b) *Let $c = \tau/6$. Then (M, g) is locally isometric to a modified Riemannian extension $g_{\nabla,\Phi,c} = \tfrac{\tau}{6} \cdot \iota \operatorname{Id} \circ \iota \operatorname{Id} + g_\nabla + \tfrac{24}{\tau}\sigma^* \nabla \rho_s.$*

If (M, g) is as in Theorem 3.9 (2), then $\mathcal{W}^+$ has a distinguished eigenvalue with a timelike associated eigenspace. The normalized eigenvector Ω defines an almost para-Hermitian structure (g, J_+), which is no longer integrable. Work of Cortés-Ayaso, Díaz-Ramos, and García-Río [59] shows that any such structure is symplectic and that (g, J_+) is an almost para-Kähler structure of constant para-holomorphic sectional curvature.

Example 3.10   The first examples of non-Ricci flat Type II Osserman metrics were given by Díaz-Ramos, García-Río, and Vázquez-Lorenzo [70]. For $0 \neq k \in \mathbb{R}$ and $f = f(x^4)$, let:

$$g = 2(dx^1 \circ dx^3 + dx^2 \circ dx^4) + (4k(x^1)^2 - \tfrac{1}{4k}f(x^4)^2)dx^3 \circ dx^3$$
$$+4k(x^2)^2 dx^4 \circ dx^4 + 2(4kx^1x^2 + x^2 f(x^4) - \tfrac{1}{4k}f'(x^4))dx^3 \circ dx^4 . \tag{3.1.b}$$

Let $^\nabla\Gamma_{21}{}^2 = {}^\nabla\Gamma_{12}{}^2 = -\tfrac{1}{2}f(x^2)$ be the Christoffel symbols of an affine connection ∇ on $\mathbb{R}^2$. One shows that the metric of Equation (3.1.b) agrees with the resulting modified Riemannian extension by computing:

$$^\nabla\rho = -\tfrac{1}{4}f(x^2)^2 dx^1 \otimes dx^1 - \tfrac{1}{2}f'(x^2)dx^1 \otimes dx^2,$$
$$^\nabla\rho_s = -\tfrac{1}{4}f(x^2)^2 dx^1 \circ dx^1 - \tfrac{1}{2}f'(x^2)dx^1 \circ dx^2,$$
$$g = 4k \cdot \iota \operatorname{Id} \circ \iota \operatorname{Id} + g_\nabla + \tfrac{1}{k}\sigma^* \,^\nabla\rho_s.$$

Ricci flat self-dual Walker metrics may be either Type II or Type III Osserman metrics. We apply the results of Section 2.3 and of Section 2.5 to describe these metrics in terms of deformed Riemannian extensions as follows:

Theorem 3.11   *A 4-dimensional Ricci flat self-dual Walker manifold (M, g) is locally isometric to a deformed Riemannian extension $g_{\nabla,\Phi}$ of an affine surface (Σ, ∇), where ∇ has a skew-symmetric Ricci tensor, and where Φ is an arbitrary symmetric $(0, 2)$-tensor field on Σ. The manifold (M, g) is Type II if and only if (Σ, ∇) is flat. The manifold (M, g) is Type III if and only if ∇ is not flat. The manifold (M, g) is flat if and only if (Σ, ∇) is flat and in a system of flat coordinates on Σ, Φ satisfies*
$$\partial_{x^2}\partial_{x^2}\Phi_{11} - 2\partial_{x^1}\partial_{x^2}\Phi_{12} + \partial_{x^1}\partial_{x^1}\Phi_{22} = 0.$$

García-Río, Vázquez-Abal, and Vázquez-Lorenzo [91] also constructed Osserman metrics with 3-step nilpotent Jacobi operators. Although such metrics are neither locally symmetric nor locally homogeneous, Theorem 3.11 provides many examples. Recently, Derdzinski [67] has constructed Type III Osserman metrics with non-nilpotent Jacobi operators—see also the work of Chudecki and Przanowski [55, 56].

Blažić, Bokan, and Rakić [18] showed that Type III Osserman metrics are foliated by degenerate surfaces that do not have to be parallel. Derdzinski [67] exhibited Type III Osserman metrics whose Jacobi operators are non-nilpotent. Any Type III Osserman 4-dimensional manifold (M, g) is locally the total space of an affine plane bundle over a surface, endowed with a distinguished "nonlinear connection" in the form of a horizontal distribution $\mathcal{H}$, transverse to the vertical distribution $\mathcal{V}$ of the bundle, both consisting of g-null vectors.

Null Jordan–Osserman Manifolds

Complex and para-complex space forms have many similarities. For instance, both have the same eigenvalue structure for their Jacobi operators. Moreover, all the first-, second-, and third-order scalar curvature invariants are exactly the same for both curvature models. The null Jordan–Osserman models of signature $(2, 2)$ are classified in Theorem 3.5. This gives rise to the following geometrical classification by García-Río *et al.* [86], which provides a geometric criterion to distinguish between para-complex and complex geometries:

Theorem 3.12 *Let (M, g) be a connected pseudo-Riemannian manifold of neutral signature $(2, 2)$. Then (M, g) is null Jordan–Osserman if and only if either it has constant sectional curvature or it is locally a complex space form.*

3.2 PARA-KÄHLER MANIFOLDS OF CONSTANT PARA-HOLOMORPHIC SECTIONAL CURVATURE

Recall from Section 1.10 that a para-Kähler manifold (M, g, J_+) is said to be of constant para-holomorphic sectional curvature c if the sectional curvature of any para-holomorphic plane is equal to c. The curvature tensor of any para-Kähler manifold of constant para-holomorphic sectional curvature is completely determined by the metric and the para-complex structure. If $\dim(M) \geq 6$, then c is constant and the curvature takes the form:

$$
\begin{aligned}
\mathcal{R}(X, Y)Z \;=\; & \tfrac{c}{4}\{g(Y, Z)X - g(X, Z)Y - g(J_+Y, Z)J_+X \\
& + g(J_+X, Z)J_+Y + 2g(J_+X, Y)J_+Z\}.
\end{aligned}
$$

Bonome *et al.* [27] established the following analogue of Lemma 1.47:

Lemma 3.13 *Let (M, g, J_+) be a para-Kähler manifold. The following conditions are equivalent:*

1. *(M, g, J_+) has constant para-holomorphic sectional curvature.*

2. *$\mathcal{J}(X)J_+X$ is proportional to J_+X, for all non-null vector fields X.*

3. *$\mathcal{J}(U)J_+U = 0$ for all null vector fields U.*

If (M, ∇) is an affine manifold, let

$$
g_{\nabla,c} := c\iota \,\mathrm{Id}\, \circ\iota \,\mathrm{Id} + g_\nabla \quad \text{on } T^*M.
$$

Let Ω be the canonical symplectic structure on T^*M, which is characterized invariantly in Equation (1.1.h) by the identity:

$$
\Omega(X^C, Y^C) = \iota[X, Y].
$$

Define an endomorphism J_+ of the tangent bundle of T^*M by the identity

$$\Omega(X, Y) = g_{\nabla,c}(X, J_+Y).$$
(3.2.a)

We use Equation (1.1.f), Equation (1.7.b), and Equation (3.2.a) to see that in a system of canonical coordinates for T^*M we have that:

$$\Omega = dx_{i'} \wedge dx^i,$$
$$g_{\nabla,c} = 2\, dx^i \circ dx_{i'} + \left(c x_{i'} x_{j'} - 2x_{k'}\, {}^\nabla\Gamma_{ij}{}^k(x)\right) dx^i \circ dx^j,$$
$$J_+(\partial_{x_i}) = \partial_{x_i} - \left\{c\, x_{i'} x_{j'} - 2x_{k'}\, {}^\nabla\Gamma_{ij}{}^k\right\} \partial_{x_{j'}} \text{ and } J_+(\partial_{x_{i'}}) = -\partial_{x_{i'}}.$$
(3.2.b)

Modified Riemannian extensions of flat connections have a special significance and provide local models for para-Kähler manifolds of constant para-holomorphic sectional curvature c.

Theorem 3.14   *Adopt the notation established above.*

1. *J_+ defines an almost para-complex structure on $(T^*M, g_{\nabla,c})$.*

2. *J_+ is integrable if and only if (M, ∇) is flat.*

3. *If (M, ∇) is flat, then $(T^*M, g_{\nabla,c}, J_+)$ is a para-Kähler manifold of constant para-holomorphic sectional curvature c.*

Proof. The fact that $J_+^2 = \mathrm{Id}$ and $J_+^* g_{\nabla,c} = -g_{\nabla,c}$ follows directly from Equation (3.2.b); Assertion (1) follows. Let $P \in M$. Choose local coordinates on M, using Lemma 1.6, so that ${}^\nabla\Gamma(P) = 0$. Let $\tilde{Q} \in \sigma^{-1}(P)$. We use Equation (3.2.b) to establish Assertion (2) by computing at $\tilde{Q}$ that:

$$J_+[J_+\partial_{x_i}, \partial_{x_j}] = 2x_{b'} \partial_{x_j}\, {}^\nabla\Gamma_{ia}{}^b \partial_{x_{a'}},$$

$$J_+[\partial_{x_i}, J_+\partial_{x_j}] = -2x_{b'} \partial_{x_i}\, {}^\nabla\Gamma_{ja}{}^b \partial_{x_{a'}},$$

$$[J_+\partial_{x_i}, J_+\partial_{x_j}] = \{2x_{b'} \partial_{x_i}\, {}^\nabla\Gamma_{ja}{}^b - 2x_{b'} \partial_{x_j}\, {}^\nabla\Gamma_{ia}{}^b\} \partial_{x_{a'}}$$

$$+ \{x_{i'} x_{a'} \partial_{x_{a'}}(x_{j'} x_{c'}) - x_{j'} x_{a'} \partial_{x_{a'}}(x_{i'} x_{c'})\} \partial_{x_{c'}}, \quad \text{so}$$

$$N_{J_+}(\partial_{x_i}, \partial_{x_j})(\tilde{Q}) = 4x_{b'}\, {}^\nabla R_{ija}{}^b(P) \partial_{x_{a'}}.$$

Let (M, ∇) be flat. As J_+ is integrable and Ω is closed, $(T^*M, g_{\nabla,c}, J_+)$ is a para-Kähler manifold. We complete the proof by showing that $(T^*M, g_{\nabla,c}, J_+)$ has constant para-holomorphic sectional curvature c. We do not sum over repeated indices in what follows. By Lemma 1.31, the non-zero components of the curvature operator of $(T^*M, g_{\nabla,c})$ are given, up to the usual symmetries, by:

$$g\nabla,c R_{ij\alpha}{}^i = \frac{c^2}{4}x_{j'}x_{\alpha'}, \qquad g\nabla,c R_{i'ii}{}^i = -c, \qquad g\nabla,c R_{i'ik}{}^k = \frac{-c}{2},$$

$$g\nabla,c R_{i'ji}{}^j = \frac{-c}{2}, \qquad g\nabla,c R_{i'ii}{}^{i'} = c^2 x_{i'}^2, \qquad g\nabla,c R_{i'ik}{}^{k'} = \frac{c^2}{2}x_{k'}^2,$$

$$g\nabla,c R_{i'ii}{}^{h'} = \frac{3c^2}{4}x_{i'}x_{h'}, \qquad g\nabla,c R_{i'ik}{}^{i'} = \frac{3c^2}{4}x_{i'}x_{k'}, \qquad g\nabla,c R_{i'ik}{}^{h'} = \frac{c^2}{2}x_{k'}x_{h'},$$

$$g\nabla,c R_{i'ji}{}^{i'} = \frac{c^2}{2}x_{i'}x_{j'}, \qquad g\nabla,c R_{i'\alpha i}{}^{h'} = \frac{c^2}{4}x_{\alpha'}x_{h'}, \qquad g\nabla,c R_{i'\alpha k}{}^{i'} = \frac{c^2}{4}x_{\alpha'}x_{k'},$$

$$g\nabla,c R_{iji}{}^{\alpha'} = \frac{-c^2}{4}x_{j'}x_{\alpha'}, \qquad g\nabla,c R_{i'ii'}{}^{i'} = c, \qquad g\nabla,c R_{i'ik'}{}^{k'} = \frac{c}{2},$$

$$g\nabla,c R_{i'jj'}{}^{i'} = \frac{c}{2}.$$

Let $X = (X^i \partial_{x_i} + X^{i'} \partial_{x_{i'}})$ and let $\varepsilon_X = g_{\nabla,c}(X, X)$. We now sum over repeated indices i and j to compute:

$$g\nabla,c \mathcal{R}(J_+ X, X)X = -\varepsilon_X \left(X^i \, g\nabla,c R_{i'ii}{}^i \, \partial_{x_i} + 2X^i \, g\nabla,c R_{i'ji}{}^{i'} \, \partial_{x^{j'}} \right)$$

$$+\varepsilon_X \left(X^i \, g\nabla,c R_{i'ii}{}^{i'} \, \partial_{x^{i'}} + X^{i'} \, g\nabla,c R_{i'ii}{}^i \, \partial_{x^{i'}} \right)$$

$$= c\,\varepsilon_X \left(X^i \, \partial_{x_i} - X^i c\, x_{i'}x_{j'} \, \partial_{x^{j'}} - X^{i'} \, \partial_{x^{i'}} \right)$$

$$= c\,\varepsilon_X \left(X^i \, J_+ \, \partial_{x_i} + X^{i'} \, J_+ \, \partial_{x^{i'}} \right).$$

Therefore, we get $g\nabla,c \mathcal{J}(X)J_+ X = c\, g_{\nabla,c}(X, X) J_+ X$. This shows that $(T^*M, g_{\nabla,c}, J_+)$ has constant para-holomorphic sectional curvature c. Since para-Kähler manifolds of constant para-holomorphic sectional curvature have a unique local-isometry type, Assertion (3) follows. $\square$

3.3   HIGHER-DIMENSIONAL OSSERMAN METRICS

In this section, we shall use Theorem 3.14 to construct new examples of Osserman manifolds by perturbing the para-complex space forms. We shall be considering affine base manifolds which are equipped with a non-flat affine Osserman connection. Since we are interested in constructing Osserman metrics as Riemannian extensions of an affine Osserman connection, for any possible Riemannian extension $g_{\nabla,\Phi,c}$ the symmetric $(0, 2)$-tensor field Φ must vanish identically by Lemma 1.39. We take $c = 1$ and consider the corresponding modified Riemannian extension $g_{\nabla,1}$.

Theorem 3.15   *Let (M, ∇) be an affine manifold of dimension $\bar{m}$.*

1. *If (M, ∇) is affine Osserman at $P \in M$, then the modified Riemannian extension $g_{\nabla,1} = \iota\, \mathrm{Id} \circ\iota\, \mathrm{Id} + g_\nabla$ on T^*M is Osserman at any $\tilde{Q} \in \sigma^{-1}(P)$. The eigenvalues of the Jacobi operators on the unit pseudo-sphere bundles $S^\pm(T_{\tilde{Q}}T^*M, g_{\nabla,1})$ are $\pm(0, 1, \frac{1}{4})$ with multiplicities $(1, 1, 2\bar{m} - 2)$, respectively.*

2. *If (M, ∇) is globally affine Osserman, then $(T^*M, g_{\nabla,1})$ is globally Osserman.*

Before beginning the proof of Theorem 3.15, we first establish some preliminary results. Let (M, ∇) be an affine manifold. Fix a point P of M. We use Lemma 1.6 to choose local coordinates $(x^1, \ldots, x^{\bar{m}})$ on M so that $\Gamma(P) = 0$ and let $^0\nabla$ be the flat connection, $^0\nabla_{\partial_{x_i}} \partial_{x_j} = 0$. We let g_0 be the Riemannian extension defined by $^0\nabla$:

$$g_0 = g_{0\nabla} = 2dx^i \circ dx_{i'} + x_{i'}x_{j'}dx^i \circ dx^j \,.$$

The auxiliary metric g_0 on T^*M is not invariantly defined but depends on the coordinates chosen; by Theorem 3.14, (T^*M, g_0, J_{+,g_0}) has constant para-holomorphic sectional curvature and thus, is Osserman.

Lemma 3.16 *Adopt the notation established above. Let ^{g_0}R be the curvature operator of g_0, let $^{g_{\nabla,1}}R$ be the curvature operator of $g_{\nabla,1}$, and let $^2R := {}^{g_{\nabla,1}}R - {}^{g_0}R$ define the Jacobi operator ^2J. If ξ is a tangent vector on T^*M, let $a = \sigma_*\xi$ be the corresponding tangent vector on M. Relative to the natural coordinate frame $\{\partial_{x_1}, \ldots, \partial_{x_{\bar{m}}}, \partial_{x^{1'}}, \ldots, \partial_{x^{\bar{m}'}}\}$ we have:*

$$^2J(\xi) = \begin{pmatrix} {}^\nabla J(a) & 0 \\ \star & {}^t({}^\nabla J(a)) \end{pmatrix} \,.$$

Proof. We write the curvature operator of $g_{\nabla,1}$ as

$$^{g_{\nabla,1}}R = {}^{g_0}R + {}^{g_\nabla}R + {}^\mathcal{E}R,$$

where $^\mathcal{E}R$ is an additional term measuring the interactions among the metrics above. It follows from Lemma 2.14 that the Jacobi operators corresponding to the Riemannian extension take a similar form:

$$^{g_0}J(\xi) = \begin{pmatrix} 0 & 0 \\ \star & 0 \end{pmatrix} \text{ and } {}^{g_\nabla}J(\xi) = \begin{pmatrix} {}^\nabla J(a) & 0 \\ \star & {}^t({}^\nabla J(a)) \end{pmatrix} \,.$$

We will complete the proof by showing that the interaction terms given by $^\mathcal{E}R$ have Jacobi operators of the form

$$^\mathcal{E}J(\xi) = \begin{pmatrix} 0 & 0 \\ \star & 0 \end{pmatrix} \,.$$

It is enough to show that the following components of $^\mathcal{E}R$ vanish:

$$\left\{ {}^\mathcal{E}R_{ijk}{}^\ell, \; {}^\mathcal{E}R_{ijk'}{}^\ell, \; {}^\mathcal{E}R_{i'jk}{}^\ell, \; {}^\mathcal{E}R_{i'jk'}{}^\ell, \; {}^\mathcal{E}R_{ij'k'}{}^\ell, \; {}^\mathcal{E}R_{ij'k'}{}^\ell, \right.$$
$$\left. {}^\mathcal{E}R_{i'j'k}{}^\ell, \; {}^\mathcal{E}R_{i'j'k'}{}^\ell, \; {}^\mathcal{E}R_{ij'k}{}^{\ell'}, \; {}^\mathcal{E}R_{ij'k'}{}^{\ell'}, \; {}^\mathcal{E}R_{i'j'k}{}^{\ell'}, \; {}^\mathcal{E}R_{i'j'k'}{}^{\ell'} \right\} \,.$$

We have $^\mathcal{E}R = {}^{g_{\nabla,1}}R - {}^{g_0}R - {}^{g_\nabla}R$. The metrics $g_{\nabla,1}$ and g_0 are Walker metrics. By Lemma 1.31, the possible non-zero components of $^\mathcal{E}R$ are:

$$\left\{ {}^\mathcal{E}R_{ij'k}{}^\ell, \; {}^\mathcal{E}R_{ij'k'}{}^{\ell'}, \; {}^\mathcal{E}R_{ijk}{}^\ell, \; {}^\mathcal{E}R_{ij'k}{}^{\ell'} \right\} \,.$$

We now show that the above four components vanish identically. By Lemma 1.31,

$$^{g_{\nabla,1}}R_{ij'k}{}^\ell = {}^{g_0}R_{ij'k}{}^\ell = \tfrac{1}{2}\{\delta_i^\ell\delta_j^k + \delta_i^j\delta_k^\ell\}, \text{ and } {}^{g_\nabla}R_{ij'k}{}^\ell = 0, \text{ so } {}^{\mathcal{E}}R_{ij'k}{}^\ell = 0,$$

$$^{g_{\nabla,1}}R_{ij'k'}{}^{\ell'} = {}^{g_0}R_{ij'k'}{}^{\ell'} = -\tfrac{1}{2}\{\delta_i^k\delta_j^\ell + \delta_i^j\delta_k^\ell\}, \text{ and } {}^{g_\nabla}R_{ij'k'}{}^{\ell'} = 0, \text{ so } {}^{\mathcal{E}}R_{ij'k'}{}^{\ell'} = 0.$$

Consider now the terms $^{\mathcal{E}}R_{ijk}{}^\ell$. Once again, using Lemma 1.31, one has

$$^{g_{\nabla,1}}R_{ijk}{}^\ell = -\partial_{x_j}{}^\nabla\Gamma_{ik}{}^\ell + \partial_{x_i}{}^\nabla\Gamma_{jk}{}^\ell$$
$$+\tfrac{1}{4}\{(\partial_{x_{s'}}(x_{j'}x_{k'}) - 2{}^\nabla\Gamma_{jk}{}^s)(\partial_{x_{\ell'}}(x_{i'}x_{s'}) - 2{}^\nabla\Gamma_{is}{}^\ell)$$
$$-(\partial_{x_{s'}}(x_{i'}x_{k'}) - 2{}^\nabla\Gamma_{ik}{}^s)(\partial_{x_{\ell'}}(x_{j'}x_{s'}) - 2{}^\nabla\Gamma_{js}{}^\ell)\},$$

$$^{g_0}R_{ijk}{}^\ell = \tfrac{1}{4}\{\partial_{x_{s'}}(x_{j'}x_{k'})\partial_{x_{\ell'}}(x_{i'}x_{s'}) - \partial_{x_{s'}}(x_{i'}x_{k'})\partial_{x_{\ell'}}(x_{j'}x_{s'})\},$$

$$^{g_\nabla}R_{ijk}{}^\ell = -\partial_{x_j}{}^\nabla\Gamma_{ik}{}^\ell + \partial_{x_i}{}^\nabla\Gamma_{jk}{}^\ell + {}^\nabla\Gamma_{jk}{}^s{}^\nabla\Gamma_{is}{}^\ell - {}^\nabla\Gamma_{ik}{}^s{}^\nabla\Gamma_{js}{}^\ell.$$

We now use the fact that $^\nabla\Gamma(P) = 0$ to see that:

$$^{g_{\nabla,1}}R_{ijk}{}^\ell = {}^{g_0}R_{ijk}{}^\ell + {}^{g_\nabla}R_{ijk}{}^\ell.$$

This implies that $^{\mathcal{E}}R_{ijk}{}^\ell = 0$. Finally, we consider those terms of the form $^{\mathcal{E}}R_{ij'k}{}^{\ell'}$. We use Lemma 1.31 to see that:

$$^{g_{\nabla,1}}R_{ij'k}{}^{\ell'} = -\partial_{x_\ell}{}^\nabla\Gamma_{ik}{}^j + \partial_{x_k}{}^\nabla\Gamma_{i\ell}{}^j$$
$$-2{}^\nabla\Gamma_{\ell s}{}^j{}^\nabla\Gamma_{ik}{}^s + {}^\nabla\Gamma_{ik}{}^s\partial_{x_{j'}}(x_{\ell'}x_{s'}) + {}^\nabla\Gamma_{\ell s}{}^r\partial_{x_{j'}}(x_{r'}\partial_{x_{s'}}(x_{i'}x_{k'}))$$
$$+\tfrac{1}{4}\{(\partial_{x_{s'}}(x_{i'}x_{k'}) - 2{}^\nabla\Gamma_{ik}{}^s)(\partial_{x_{j'}}(x_{s'}x_{\ell'}) - 2{}^\nabla\Gamma_{s\ell}{}^j)$$
$$+(\partial_{x_{s'}}(x_{i'}x_{\ell'}) - 2{}^\nabla\Gamma_{i\ell}{}^s)(\partial_{x_{j'}}(x_{s'}x_{k'}) - 2{}^\nabla\Gamma_{sk}{}^j)$$
$$-2\partial_{x_{j'}}(x_{\ell'}x_{s'}\partial_{x_{s'}}(x_{i'}x_{k'}))\},$$

$$^{g_0}R_{ij'k}{}^{\ell'} = \tfrac{1}{4}\{\partial_{x_{s'}}(x_{i'}x_{k'})\partial_{x_{j'}}(x_{s'}x_{\ell'}) + \partial_{x_{s'}}(x_{i'}x_{\ell'})\partial_{x_{j'}}(x_{s'}x_{k'})$$
$$-2\partial_{x_{j'}}(x_{\ell'}x_{s'}\partial_{x_{s'}}(x_{i'}x_{k'}))\},$$

$$^{g_\nabla}R_{ij'k}{}^{\ell'} = -\partial_{x_\ell}{}^\nabla\Gamma_{ik}{}^j + \partial_{x_k}{}^\nabla\Gamma_{i\ell}{}^j - {}^\nabla\Gamma_{ik}{}^s{}^\nabla\Gamma_{s\ell}{}^j + {}^\nabla\Gamma_{i\ell}{}^s{}^\nabla\Gamma_{sk}{}^j.$$

Since $^\nabla\Gamma(P) = 0$, $^{g_{\nabla,1}}R_{ij'k}{}^{\ell'} = {}^{g_0}R_{ij'k}{}^{\ell'} + {}^{g_\nabla}R_{ij'k}{}^{\ell'}$, so $^{\mathcal{E}}R_{ij'k}{}^{\ell'} = 0$ as desired. $\qquad\square$

We continue the discussion; we will use the underlying almost para-Kähler structure J_+ of $(T^*M, g_{\nabla,1})$, which is given in Equation (3.2.b). Let $\xi \in S^+(T_{\tilde{Q}}T^*M, g_{\nabla,1})$. Set

$$\xi_1 := J_+\xi \in S^-(T_{\tilde{Q}}T^*M, g_{\nabla,1}).$$

Let $E_\lambda(\xi)$ (resp. $E_\lambda(\xi_1)$) be the eigenspaces of $^{g_0}\mathcal{J}(\xi)$ (resp. $^{g_0}\mathcal{J}(\xi_1)$) for the eigenvalue $\lambda \in \{0, 1, \tfrac{1}{4}\}$ (resp. for $\lambda \in \{0, -1, -\tfrac{1}{4}\}$). Set

$$E_0(\xi) = \xi \cdot \mathbb{R} = E_{-1}(\xi_1), \qquad\qquad E_1(\xi) = \xi_1 \cdot \mathbb{R} = E_0(\xi_1),$$
$$E_{\frac{1}{4}}(\xi) = \{E_0(\xi) \oplus E_1(\xi)\}^\perp = \{E_{-1}(\xi_1) \oplus E_0(\xi_1)\}^\perp = E_{-\frac{1}{4}}(\xi_1), \qquad (3.3.a)$$
$$\mathcal{S}(\xi) := \mathfrak{D} \cap E_{\frac{1}{4}}(\xi), \qquad\qquad \mathcal{U}(\xi) := E_0(\xi) \oplus \mathfrak{D},$$

where $\mathfrak{D} = \ker\{\sigma_*\}$ is the null parallel vertical plane field on $(T^*M, g_{\nabla,1})$. We then have $T_{\tilde{Q}} T^*M = E_0(\xi) \oplus E_1(\xi) \oplus E_{\frac{1}{4}}(\xi)$.

Lemma 3.17   *Let (M, ∇) be an affine manifold. Let $\tilde{Q} \in T^*M$. Let $\xi \in S^+(T_{\tilde{Q}} T^*M, g_{\nabla,1})$.*

1. *$\mathfrak{D} = (\xi_1 - \xi) \cdot \mathbb{R} + S(\xi)$.*

2. *${}^2 \mathcal{J}(\xi)\mathfrak{D} \subset S(\xi)$.*

3. *${}^{g_0} \mathcal{J}(\xi) \mathcal{U}(\xi) \subset \mathcal{U}(\xi)$ and ${}^2 \mathcal{J}(\xi) \mathcal{U}(\xi) \subset \mathcal{U}(\xi)$.*

Proof. First of all, observe that Equation (3.2.b) implies $\xi_1 - \xi \in \mathfrak{D}$. Choose an orthonormal basis for $E_{\frac{1}{4}}(\xi)$ of the form $\{e_1^+, \ldots, e_{m-1}^+, J_+ e_1^+, \ldots, J_+ e_{m-1}^+\}$, where the e_i^+ are spacelike and the $J_+ e_i^+$ are timelike. Then, Assertion (1) follows by noting that we have the following basis for $\mathfrak{D}$:

$$\{\xi - \xi_1, e_1^+ - J_+ e_1^+, \ldots, e_{m-1}^+ - J_+ e_{m-1}^+\}.$$

Suppose now that Assertion (2) fails and choose $\eta \in \mathfrak{D}$ so that ${}^2 \mathcal{J}(\xi)\eta \notin S(\xi)$. Then it follows by Lemma 3.16 that ${}^2 \mathcal{J}(\xi)\eta \in \mathfrak{D}$. By Assertion (1), there exists $c \neq 0$ so that

$$ {}^2 \mathcal{J}(\xi)\eta = c(\xi - \xi_1) + \eta_1 \quad \text{for} \quad \eta_1 \in S(\xi). $$

Thus, $c\xi \in E_1(\xi) + \text{Range}\{{}^2 \mathcal{J}(\xi)\} + E_{\frac{1}{4}}(\xi) \subset E_0(\xi)^\perp$, which is false; this contradiction establishes Assertion (2). To prove Assertion (3), express:

$$\mathcal{U}(\xi) = E_0(\xi) \oplus \mathfrak{D} = \xi \cdot \mathbb{R} \oplus (\xi_1 - \xi) \cdot \mathbb{R} \oplus S(\xi) = \xi \cdot \mathbb{R} \oplus \xi_1 \cdot \mathbb{R} \oplus S(\xi). \quad (3.3.b)$$

Since ${}^{g_0} \mathcal{J}(\xi)\xi = 0$, ${}^{g_0} \mathcal{J}(\xi)\xi_1 = \xi_1$, and $S(\xi) \subset E_{\frac{1}{4}}(\xi)$, ${}^{g_0} \mathcal{J}(\xi)$ preserves $\mathcal{U}(\xi)$. Analogously, since ${}^2 \mathcal{J}(\xi)\xi = 0$ and ${}^2 \mathcal{J}(\xi)\mathfrak{D} \subset \mathfrak{D}$, one has that ${}^2 \mathcal{J}(\xi)$ preserves $\mathcal{U}(\xi)$ as well. $\square$

We can now examine the eigenvalue structure:

Lemma 3.18   *Let (M, ∇) be an affine manifold. Let $\tilde{Q} \in T^*M$. Let $\xi \in S^+(T_{\tilde{Q}} T^*M, g_{\nabla,1})$. Assume (M, ∇) is affine Osserman at $P = \sigma(\tilde{Q})$. If there is $0 \neq \eta \in T_{\tilde{Q}} T^*M \otimes_{\mathbb{R}} \mathbb{C}$ with ${}^{g_{\nabla,1}} \mathcal{J}(\xi)\eta = \mu\eta$, then:*

1. *If $\eta \notin \mathcal{U}(\xi) \otimes_{\mathbb{R}} \mathbb{C}$, then $\mu = \frac{1}{4}$.*

2. *If $\eta \in \mathcal{U}(\xi) \otimes_{\mathbb{R}} \mathbb{C}$ and if $\eta \notin S(\xi) \otimes_{\mathbb{R}} \mathbb{C}$, then $\mu = 0$ or $\mu = 1$.*

3. *If $\eta \in S(\xi) \otimes_{\mathbb{R}} \mathbb{C}$, then $\mu = \frac{1}{4}$.*

4. *$\text{Spec}\{{}^{g_{\nabla,1}} \mathcal{J}(\xi)\} \subset \{0, 1, \frac{1}{4}\}$.*

Proof. Let (M, ∇) be affine Osserman at P. This means that $^\nabla \mathcal{J}(\xi)$ is nilpotent at P. Thus, by Lemma 3.16, $^2\mathcal{J}(\xi)$ is also nilpotent at any point $\tilde{Q}$ in the fiber over P. By Assertion (3) in Lemma 3.17, $\mathcal{U}(\xi)$ is preserved by both $^{g_0}\mathcal{J}(\xi)$ and $^2\mathcal{J}(\xi)$. Let $^{g_0}\tilde{\mathcal{J}}(\xi)$, $^2\tilde{\mathcal{J}}(\xi)$, and $^{g_{\nabla,1}}\tilde{\mathcal{J}}(\xi) = {}^{g_0}\tilde{\mathcal{J}}(\xi) + {}^2\tilde{\mathcal{J}}(\xi)$ be the induced operators on the quotient space:

$$\mathcal{V}(\xi) := \{T_{\tilde{Q}}T^*M/\mathcal{U}(\xi)\} \otimes_{\mathbb{R}} \mathbb{C}.$$

If $\eta \notin \mathcal{U}(\xi) \otimes_{\mathbb{R}} \mathbb{C}$, then the projection $\tilde{\eta} \in \mathcal{V}(\xi)$ of η is different from zero and we have that $^{g_{\nabla,1}}\tilde{\mathcal{J}}(\xi)\tilde{\eta} = \mu\tilde{\eta}$. By Equation (3.3.b), one now has that

$$\mathcal{V}(\xi) = \{E_{\frac{1}{4}}(\xi)/\mathcal{S}(\xi)\} \otimes_{\mathbb{R}} \mathbb{C}.$$

Consequently, $^{g_0}\tilde{\mathcal{J}}(\xi) = \frac{1}{4}\,\mathrm{Id}$. Since $^2\tilde{\mathcal{J}}(\xi)$ is nilpotent and $^{g_{\nabla,1}}\tilde{\mathcal{J}}(\xi) = \frac{1}{4}\,\mathrm{Id} + {}^2\tilde{\mathcal{J}}(\xi)$, one obtains that $^{g_{\nabla,1}}\tilde{\mathcal{J}}(\xi)$ has only the eigenvalue $\frac{1}{4}$. Thus, $\mu = \frac{1}{4}$, which establishes Assertion (1).

In order to prove Assertion (2), suppose that there exists $0 \neq \eta \in \mathcal{U}(\xi) \otimes_{\mathbb{R}} \mathbb{C}$ such that $\eta \notin \mathcal{S}(\xi) \otimes_{\mathbb{R}} \mathbb{C}$ and $^{g_{\nabla,1}}\mathcal{J}(\xi)\eta = \mu\eta$. Note that $\mathcal{S}(\xi)$ is preserved by the Jacobi operators $^{g_0}\mathcal{J}(\xi)$ and $^2\mathcal{J}(\xi)$. We apply Lemma 3.17 and consider the operators $^{g_0}\tilde{\mathcal{J}}(\xi)$, $^2\tilde{\mathcal{J}}(\xi)$, and $^{g_{\nabla,1}}\tilde{\mathcal{J}}(\xi) = {}^{g_0}\tilde{\mathcal{J}}(\xi) + {}^2\tilde{\mathcal{J}}(\xi)$ on the quotient space

$$\mathcal{Y}(\xi) := \{\mathcal{U}(\xi)/\mathcal{S}(\xi)\} \otimes_{\mathbb{R}} \mathbb{C}.$$

Since $\eta \notin \mathcal{S}(\xi) \otimes_{\mathbb{R}} \mathbb{C}$, its projection $\tilde{\eta} \neq 0$ and μ is an eigenvalue of $^{g_{\nabla,1}}\tilde{\mathcal{J}}(\xi)$. Now Equation (3.3.b) shows that $\mathcal{Y}(\xi) = \tilde{\xi} \cdot \mathbb{R} \oplus \tilde{\xi}_1 \cdot \mathbb{R}$. By Lemma 3.17,

$$^2\mathcal{J}(\xi)\xi = 0 \quad \text{and} \quad ^2\mathcal{J}(\xi)\xi_1 = {}^2\mathcal{J}(\xi)(\xi_1 - \xi) \in \mathcal{S}(\xi).$$

Consequently, $^2\tilde{\mathcal{J}}(\xi) = 0$. Since $^{g_0}\tilde{\mathcal{J}}(\xi)\tilde{\xi} = 0$ and $^{g_0}\tilde{\mathcal{J}}(\xi)\tilde{\xi}_1 = \tilde{\xi}_1$ we have that $^{g_{\nabla,1}}\tilde{\mathcal{J}}(\xi)\tilde{\xi} = 0$ and $^{g_{\nabla,1}}\tilde{\mathcal{J}}(\xi)\tilde{\xi}_1 = \tilde{\xi}_1$, which shows that $\mu \in \{0, 1\}$, and Assertion (2) follows.

To prove Assertion (3), we note that $^{g_0}\mathcal{J}(\xi) = \frac{1}{4}\,\mathrm{Id}$ on $\mathcal{S}(\xi)$ and that $^2\mathcal{J}(\xi)$ is nilpotent and preserves $\mathcal{S}(\xi)$. Assertion (4) follows from Assertions (1)-(3). $\square$

Proof of Theorem 3.15: Let (M, ∇) be an affine manifold which is affine Osserman at $P \in M$. Use Lemma 1.6 to choose local coordinates on M so that $^\nabla\Gamma(P) = 0$ and let $^0\nabla$ be the flat affine connection defined on a neighborhood of P whose Christoffel symbols vanish in these coordinates. Define a 1-parameter family of metrics on T^*M interpolating between $g_{\nabla,1}$ and g_0 by considering the modified Riemannian extensions associated to the connections

$$^\varepsilon\nabla := \varepsilon\nabla + (1 - \varepsilon)^0\nabla.$$

Since $^\nabla\Gamma(P) = {}^0\nabla\Gamma(P) = 0$ and since $^0\nabla\mathcal{R}(P) = 0$, the corresponding curvature operators satisfy $^\varepsilon\nabla\mathcal{R}(P) = \varepsilon \cdot {}^\nabla\mathcal{R}(P)$. Thus all the connections $^\varepsilon\nabla$ are affine Osserman at P. By Lemma 3.18,

$$\mathrm{Spec}\{^{g_\varepsilon\nabla}\mathcal{J}(\xi)\} \subset \{0, 1, \tfrac{1}{4}\}$$

for all ε and hence, the eigenvalues and their multiplicities are unchanged during perturbation defined by ε. Taking $\varepsilon = 0$ yields the desired multiplicities and establishes Assertion (1) of Theorem 3.15 for ξ spacelike. We now use results of García-Río, Kupeli, and Vázquez-Lorenzo [90] to see that spacelike Osserman implies timelike Osserman and to relate the eigenvalues and eigenvalue multiplicities on $S^+(T_{\tilde{Q}}T^*M, g_{\nabla,1})$ to the eigenvalues and eigenvalue multiplicities on $S^-(T_{\tilde{Q}}T^*M, g_{\nabla,1})$; alternatively, of course, one could simply proceed directly as well. This proves Assertion (1) of Theorem 3.15; Assertion (2) follows from Assertion (1). $\quad\square$

The computation of the curvature on the *zero section* $\mathcal{Z}(T^*M)$ is qualitatively easier than in the general setting since we may set $\vec{x}' = 0$.

Lemma 3.19 *Let (M, ∇) be an affine manifold and let $\tilde{Q} \in \mathcal{Z}(T_P^*M)$. Choose local coordinates so $^\nabla\Gamma(P) = 0$. Then:*

1. *The possibly non-zero components of $^{g\nabla}R(\tilde{Q})$ are $^{g\nabla}R_{ijk\ell'}(\tilde{Q}) = {}^\nabla R_{ijk}{}^\ell(P)$.*

2. *The non-zero components of $^{g_0}R(\tilde{Q})$ are: $^{g_0}R_{ij'i'j}(\tilde{Q}) = {}^{g_0}R_{ii'j'j}(\tilde{Q}) = -\frac{1}{2}$ for $i \neq j$ and $^{g_0}R_{ii'i'i}(\tilde{Q}) = -1$.*

3. *We have $^{g\nabla,1}R(\tilde{Q}) = {}^{g_0}R(\tilde{Q}) + {}^{g\nabla}R(\tilde{Q})$.*

Proof. Let $u = (u^1, \ldots, u^n)$ be coordinates on a pseudo-Riemannian manifold (U, h). Expand $h = h_{ab} du^a \circ du^b$. If the 1-jets of the components h_{ab} vanish at a point S of U, then differentiating the Koszul formula yields:

$$^hR_{abcd}(S) = \tfrac{1}{2}\{\partial_{u_a}\partial_{u_c}h_{bd} + \partial_{u_b}\partial_{u_d}h_{ac} - \partial_{u_a}\partial_{u_d}h_{bc} - \partial_{u_b}\partial_{u_c}h_{ad}\}(S). \qquad (3.3.c)$$

We apply this observation to the setting at hand. Let $\tilde{Q}$ be a point in the zero section of T^*M. Then the coordinates $\vec{x}'$ vanish at $\tilde{Q}$ and, since $^\nabla\Gamma(P)$ vanishes at P, the 1-jets of the coefficients of the metrics g_∇, g_0, and $g_{\nabla,1}$ vanish at $\tilde{Q}$. We establish Assertion (1) by using Equation (3.3.c) to compute:

$$
\begin{aligned}
^{g\nabla}R_{ijk\ell'}(\tilde{Q}) &= \tfrac{1}{2}\{\partial_{x_j}\partial_{x\ell'}(-2x_{h'}{}^\nabla\Gamma_{ik}{}^h) - \partial_{x_i}\partial_{x\ell'}(-2x_{h'}{}^\nabla\Gamma_{jk}{}^h)\}(\tilde{Q}) \\
&= \{\partial_{x_i}{}^\nabla\Gamma_{jk}{}^\ell - \partial_{x_j}{}^\nabla\Gamma_{ik}{}^\ell\}(P) \\
&= {}^\nabla R_{ijk}{}^\ell(P).
\end{aligned}
$$

The proof of Assertion (2) and of Assertion (3) is similar. $\quad\square$

As an immediate application of Theorem 3.15, observe that, since the product $M_1 \times M_2$ of affine Osserman manifolds $(M_1, {}^{(1)}\nabla)$ and $(M_2, {}^{(2)}\nabla)$ is still affine Osserman, the corresponding cotangent bundle $T^*(M_1 \times M_2)$ provides examples of Osserman manifolds with non-zero eigenvalues. Moreover, the Jacobi operators are not diagonalizable in general. The following result shows that there is no scarcity of examples of affine Osserman connections and, moreover, that the Jordan normal form of the Jacobi operators $^{g\nabla,1}\mathcal{J}$ can be quite complicated; it also shows that $(T^*M, g_{\nabla,1})$ need not be Jordan–Osserman:

Theorem 3.20   *Let $r \geq 2$ and let U be an $r \times r$ lower triangular matrix. There exists an affine Osserman manifold (M, ∇) of dimension $r + 1$, there exists a point $\tilde{Q}$ in the zero section of the cotangent bundle, and there exist $\xi_i \in S^+(T_{\tilde{Q}}T^*M, g_{\nabla,1})$ for $i = 1, 2$ so that $^{g\nabla,1}\mathcal{J}(\xi_1)$ is diagonalizable and so that relative to a suitable basis for $\tilde{T}_{\tilde{Q}}T^*M$,*

$$^{g\nabla,1}\mathcal{J}(\xi_2) = 0 \cdot \mathrm{Id}_1 \oplus 1 \cdot \mathrm{Id}_1 \oplus (\tfrac{1}{4} \cdot \mathrm{Id}_r + U) \oplus (\tfrac{1}{4} \cdot \mathrm{Id}_r + {}^t U).$$

Proof. Let $r \geq 2$ and let $M := \mathbb{R}^{r+1}$. Let $\{x^0, \ldots, x^r\}$ be the usual coordinates on M and $\{x_{0'}, \ldots, x_{r'}\}$ the dual fiber coordinates on T^*M. We let indices a, b, c, and d range from 1 through r and indices i, j, and k range from 0 through r. Let $U_a{}^b$ be a lower triangular matrix (i.e., $U_a{}^b = 0$ for $b \leq a$). Let $\theta = \theta(x^0) \in C^\infty(\mathbb{R})$. Define a torsion-free connection ∇ on TM with non-zero Christoffel symbols:

$$^\nabla\Gamma_{0a}{}^b = {}^\nabla\Gamma_{a0}{}^b = \theta(x^0)U_a{}^b.$$

This satisfies the hypotheses of Definition 2.3 and thus, by Theorem 2.4, (M, ∇) is geodesically complete, Ricci flat, and affine Osserman. The possibly non-zero curvatures are:

$$^\nabla R_{0a0}{}^b = \partial_{x_0}{}^\nabla\Gamma_{a0}{}^b + {}^\nabla\Gamma_{0c}{}^b{}^\nabla\Gamma_{a0}{}^c = \partial_{x_0}\theta \cdot U_a{}^b + \theta^2 \cdot U_c{}^b U_a{}^c.$$

Set $P = 0$, $\tilde{Q} = (0, 0)$ and assume $\theta(0) = 0$ and $\partial_{x_0}\theta(0) = -1$. We may then apply Lemma 3.19 to get that the possibly non-zero components of the curvature at $\tilde{Q}$ are given by $^{g\nabla,1}R(\tilde{Q}) = {}^{g0}R(\tilde{Q}) + {}^{g\nabla}R(\tilde{Q})$ and moreover

$$^{g\nabla,1}R_{ii'i'i}(\tilde{Q}) = -1, \qquad {}^{g\nabla,1}R_{ijkd'}(\tilde{Q}) = {}^\nabla R_{ijk}{}^d(P),$$

$$^{g\nabla,1}R_{ij'i'j}(\tilde{Q}) = {}^{g\nabla,1}R_{ii'j'j}(\tilde{Q}) = -\tfrac{1}{2} \quad (i \neq j).$$

Let $\xi_1 := \tfrac{1}{\sqrt{2}}(\partial_{x_1} + \partial_{x_{1'}})$. Then $^{g\nabla,1}\mathcal{J}(\xi_1) = {}^{g0}\mathcal{J}(\xi_1)$, so $^{g\nabla,1}\mathcal{J}(\xi_1)$ is diagonalizable; the curvature of ∇ plays no role. Take $\xi_2 := \tfrac{1}{\sqrt{2}}(\partial_{x_0} + \partial_{x_{0'}})$ and $J_+\xi_2 = \tfrac{1}{\sqrt{2}}(\partial_{x_0} - \partial_{x_{0'}})$. Then:

$$^{g0}\mathcal{J}(\xi_2)\xi_2 = 0, \qquad {}^{g\nabla}\mathcal{J}(\xi_2)\xi_2 = 0, \qquad {}^{g0}\mathcal{J}(\xi_2)\partial_{x_a} = \tfrac{1}{4}\partial_{x_a},$$

$$^{g\nabla}\mathcal{J}(\xi_2)\partial_{x_a} = U_a{}^b\partial_{x_b}, \qquad {}^{g0}\mathcal{J}(\xi_2)J_+\xi_2 = J_+\xi_2, \qquad {}^{g\nabla}\mathcal{J}(\xi_2)J_+\xi_2 = 0,$$

$$^{g0}\mathcal{J}(\xi_2)\partial_{x_{a'}} = \tfrac{1}{4}\partial_{x_{a'}}, \qquad {}^{g\nabla}\mathcal{J}(\xi_2)\partial_{x_{a'}} = U_b{}^a\partial_{x_{b'}}.$$

This shows that the Jacobi operator $^{g\nabla,1}\mathcal{J}(\xi_2)$ satisfies

$$^{g\nabla,1}\mathcal{J}(\xi_2) = 0 \cdot \mathrm{Id}_1 \oplus 1 \cdot \mathrm{Id}_1 \oplus (\tfrac{1}{4} \cdot \mathrm{Id}_r + U) \oplus (\tfrac{1}{4} \cdot \mathrm{Id}_r + {}^t U). \qquad \square$$

3.4   OSSERMAN METRICS WITH NON-TRIVIAL JORDAN NORMAL FORM

In Section 3.4, we will construct a family of Osserman manifolds of neutral signature whose Jacobi operators are not nilpotent and that have non-trivial Jordan normal form. Let $(\Sigma, \bar{\nabla})$ be an affine Osserman surface with non-zero curvature tensor at some point. Let

$$(M, \nabla) := (\Sigma \times \mathbb{R}^{\bar{m}-2}, \bar{\nabla} \oplus {}^{0}\nabla),$$

where ${}^{0}\nabla$ is the flat Euclidean connection on $\mathbb{R}^{\bar{m}-2}$. Since (M, ∇) is affine Osserman, Theorem 3.15 shows that $(T^*M, g_{\nabla,1})$ is Osserman and that the spectrum of the Jacobi operator on $S^+(T^*M)$ is $\{0, 1, \frac{1}{4}, \ldots, \frac{1}{4}\}$. By Theorem 2.6 (4), we may choose local coordinates (x^1, x^2) on Σ so that the possible non-zero Christoffel symbols are:

$$\bar{\nabla}\Gamma_{11}{}^{1} = -\partial_{x_1}\theta \quad \text{and} \quad \bar{\nabla}\Gamma_{22}{}^{2} = \partial_{x_2}\theta \quad \text{where} \quad \partial_{x_2}\partial_{x_1}\theta(x^1, x^2) \neq 0. \qquad (3.4.a)$$

Let $\{i, j, k, h\}$ be distinct indices ranging from 1 to $\bar{m}$ and let $\{\beta, \gamma, \delta\}$ be indices ranging from 1 to $\bar{m}$. By Lemma 1.31, the possibly non-zero components of curvature tensor of $(T^*M, g_{\nabla,1})$ are given by (do not sum):

$$g_{\nabla,1}R_{ij\gamma}{}^{i} = \tfrac{1}{4}x_{j'}x_{\gamma'}, \; j = \gamma > 2 \text{ or, otherwise } j \neq \gamma \text{ and } i > 2, \text{ or } j > 2, \text{ or } \gamma > 2,$$

$$g_{\nabla,1}R_{211}{}^{2'} = x_{2'}\partial_{x_1}\partial_{x_1}\partial_{x_2}\theta - x_{1'}\partial_{x_1}\partial_{x_2}\partial_{x_2}\theta + (x_{1'}\partial_{x_2}\theta + x_{2'}(x_{1'} + \partial_{x_1}\theta))\partial_{x_1}\partial_{x_2}\theta,$$

$$g_{\nabla,1}R_{212}{}^{1'} = x_{1'}\partial_{x_1}\partial_{x_2}\partial_{x_2}\theta - x_{2'}\partial_{x_1}\partial_{x_1}\partial_{x_2}\theta - (x_{1'}\partial_{x_2}\theta - x_{2'}(x_{1'} - \partial_{x_1}\theta))\partial_{x_1}\partial_{x_2}\theta,$$

$$g_{\nabla,1}R_{i'\beta i}{}^{\delta'} = \tfrac{1}{4}x_{\beta'}x_{\delta'}, \; i \neq \beta = \delta > 2 \text{ or, otherwise}$$
$$\beta \neq \delta \neq i \neq \beta, \text{ and } i > 2, \text{ or } \beta > 2, \text{ or } \delta > 2,$$

$$g_{\nabla,1}R_{i'\beta\gamma}{}^{i'} = \tfrac{1}{4}x_{\beta'}x_{\gamma'}, \; i \neq \beta = \gamma > 2 \text{ or, otherwise}$$
$$\beta \neq \gamma \neq i \neq \beta \text{ and } \beta \neq 2 \text{ or } \gamma \neq 2$$

$$g_{\nabla,1}R_{iji}{}^{\delta'} = -\tfrac{1}{4}x_{j'}x_{\delta'}, \; j = \delta > 2 \text{ or, otherwise } j \neq \delta \text{ and } i > 2 \text{ or } j > 2 \text{ or } \delta > 2,$$

$$g_{\nabla,1}R_{211}{}^{1} = -\tfrac{1}{4}x_{1'}x_{2'} - \partial_{x_1}\partial_{x_2}\theta, \qquad\qquad g_{\nabla,1}R_{212}{}^{2} = \tfrac{1}{4}x_{1'}x_{2'} - \partial_{x_1}\partial_{x_2}\theta,$$

$$g_{\nabla,1}R_{i11}{}^{i} = \tfrac{1}{4}x_{1'}(x_{1'} + 2\partial_{x_1}\theta), \; i > 1, \qquad g_{\nabla,1}R_{i22}{}^{i} = \tfrac{1}{4}x_{2'}(x_{2'} - 2\partial_{x_2}\theta), \; i \neq 2,$$

$$g_{\nabla,1}R_{211}{}^{1'} = x_{1'}(x_{1'} + 2\partial_{x_1}\theta)\partial_{x_1}\partial_{x_2}\theta, \qquad g_{\nabla,1}R_{212}{}^{2'} = x_{2'}(x_{2'} - 2\partial_{x_2}\theta)\partial_{x_1}\partial_{x_2}\theta,$$

$$g_{\nabla,1}R_{i11}{}^{2'} = -x_{1'}x_{i'}\partial_{x_1}\partial_{x_2}\theta, \; i > 2, \qquad g_{\nabla,1}R_{i22}{}^{1'} = x_{2'}x_{i'}\partial_{x_1}\partial_{x_2}\theta, \; i > 2,$$

$$g_{\nabla,1}R_{i12}{}^{1'} = x_{1'}x_{i'}\partial_{x_1}\partial_{x_2}\theta, \; i > 2, \qquad g_{\nabla,1}R_{i21}{}^{2'} = -x_{2'}x_{i'}\partial_{x_1}\partial_{x_2}\theta, \; i > 2,$$

$$g_{\nabla,1}R_{21k}{}^{1'} = x_{1'}x_{k'}\partial_{x_1}\partial_{x_2}\theta, \; k > 2, \qquad g_{\nabla,1}R_{21k}{}^{2'} = x_{2'}x_{k'}\partial_{x_1}\partial_{x_2}\theta, \; k > 2,$$

$$g_{\nabla,1}R_{1'11}{}^{1'} = x_{1'}(x_{1'} + 2\partial_{x_1}\theta), \qquad g_{\nabla,1}R_{1'11}{}^{2'} = \tfrac{3}{4}x_{1'}x_{2'} - \partial_{x_1}\partial_{x_2}\theta,$$

$$g_{\nabla,1}R_{2'22}{}^{2'} = x_{2'}(x_{2'} - 2\partial_{x_2}\theta), \qquad g_{\nabla,1}R_{2'22}{}^{1'} = \tfrac{3}{4}x_{1'}x_{2'} + \partial_{x_1}\partial_{x_2}\theta,$$

$$g_{\nabla,1}R_{1'12}{}^{1'} = \tfrac{3}{4}x_{1'}x_{2'} + \partial_{x_1}\partial_{x_2}\theta, \qquad g_{\nabla,1}R_{2'21}{}^{2'} = \tfrac{3}{4}x_{1'}x_{2'} - \partial_{x_1}\partial_{x_2}\theta,$$

$$g_{\nabla,1}R_{i'11}{}^{i'} = \tfrac{1}{4}x_{1'}(x_{1'} + 2\partial_{x_1}\theta), \; i > 1, \qquad g_{\nabla,1}R_{i'22}{}^{i'} = \tfrac{1}{4}x_{2'}(x_{2'} - 2\partial_{x_2}\theta), \; i \neq 2,$$

$$g_{\nabla,1}R_{i'ii}{}^{i'} = x_i^2, i > 2,$$
$$g_{\nabla,1}R_{i'i1}{}^{1'} = \tfrac{1}{2}x_{1'}(x_{1'} + 2\partial_{x_1}\theta), i > 1,$$

$$g_{\nabla,1}R_{i'i2}{}^{2'} = \tfrac{1}{2}x_{2'}(x_{2'} - 2\partial_{x_2}\theta), i \neq 2,$$
$$g_{\nabla,1}R_{i'ik}{}^{k'} = \tfrac{1}{2}x_k^2, k > 2,$$

$$g_{\nabla,1}R_{i'1i}{}^{1'} = \tfrac{1}{4}x_{1'}(x_{1'} + 2\partial_{x_1}\theta), i > 1,$$
$$g_{\nabla,1}R_{i'2i}{}^{2'} = \tfrac{1}{4}x_{2'}(x_{2'} - 2\partial_{x_2}\theta), i \neq 2,$$

$$g_{\nabla,1}R_{i'ii}{}^{h'} = \tfrac{3}{4}x_{i'}x_{h'}, i > 2 \text{ or } h > 2,$$
$$g_{\nabla,1}R_{i'ik}{}^{i'} = \tfrac{3}{4}x_{i'}x_{k'}, i > 2 \text{ or } k > 2,$$

$$g_{\nabla,1}R_{i'ik}{}^{h'} = \tfrac{1}{2}x_{k'}x_{h'},$$
$$g_{\nabla,1}R_{i'ji}{}^{i'} = \tfrac{1}{2}x_{i'}x_{j'},$$
$$g_{\nabla,1}R_{212'}{}^{2'} = -\tfrac{1}{4}x_{1'}x_{2'} + \partial_{x_1}\partial_{x_2}\theta,$$

$$g_{\nabla,1}R_{211'}{}^{1'} = \tfrac{1}{4}x_{1'}x_{2'} + \partial_{x_1}\partial_{x_2}\theta,$$
$$g_{\nabla,1}R_{211'}{}^{2'} = \tfrac{1}{4}x_{2'}(x_{2'} - 2\partial_{x_2}\theta),$$

$$g_{\nabla,1}R_{i1i'}{}^{1'} = -\tfrac{1}{4}x_{1'}(x_{1'} + 2\partial_{x_1}\theta), i > 1,$$
$$g_{\nabla,1}R_{i2i'}{}^{2'} = -\tfrac{1}{4}x_{2'}(x_{2'} - 2\partial_{x_2}\theta), i > 2,$$

$$g_{\nabla,1}R_{i'ii}{}^{i} = -1,$$
$$g_{\nabla,1}R_{i'ik}{}^{k} = -\tfrac{1}{2},$$
$$g_{\nabla,1}R_{i'ji}{}^{j} = -\tfrac{1}{2},$$

$$g_{\nabla,1}R_{i'ii}{}^{i'} = 1,$$
$$g_{\nabla,1}R_{i'ik'}{}^{k'} = \tfrac{1}{2},$$
$$g_{\nabla,1}R_{i'jj'}{}^{i'} = \tfrac{1}{2}.$$

We use the following technical result to simplify our subsequent computations:

Lemma 3.21 *Let* $(M, \nabla) = (\Sigma \times \mathbb{R}^{\bar{m}-2}, \tilde{\nabla} \oplus {}^0\nabla)$, *and let* $\xi \in S^+(T_{\tilde{P}}(T^*M))$ *for* $\tilde{P} \in T^*M$. *There exists an isometry* Θ *of* $(T^*M, g_{\nabla,1})$ *that preserves the associated almost para-complex structure* J_+ *so that:*

1. *If* $\bar{m} \geq 3$, *then* $\Theta(\tilde{P}) = (a_1, a_2, 0, \ldots, 0, b_1, b_2, b_3, 0, \ldots, 0)$.

2. *If* $\bar{m} \geq 5$, *then* $\Theta_*\xi = (c_1, c_2, c_3, c_4, 0, \ldots, 0, d_1, d_2, d_3, d_4, d_5, 0, \ldots, 0)$.

3. *Let* $\mathcal{J} := {}^{g_{\nabla,1}}\mathcal{J}_{\Theta(\tilde{P})}(\Theta_*\xi)$. *Relative to the canonical coordinate frame:*

$$\mathcal{J}\partial_{x_i} = \tfrac{1}{4}\partial_{x_i} \text{ and } \mathcal{J}\partial_{x_{i'}} = \tfrac{1}{4}\partial_{x_{i'}} \text{ for } 6 \leq i \leq \bar{m},$$
$$\mathcal{J}\,\text{Span}\{\partial_{x_i}, \partial_{x_{i'}}\}_{1 \leq i \leq 5} \subset \text{Span}\{\partial_{x_i}, \partial_{x_{i'}}\}_{1 \leq i \leq 5}.$$

Proof. Let $\bar{m} \geq 3$. Let $(x^1, \ldots, x^{\bar{m}}, x_{1'}, \ldots, x_{\bar{m}'})$ be canonical local coordinates on T^*M. Let $\tilde{P} = (a^1, a^2, \ldots, a^{\bar{m}}, \star) \in T^*M$, where $\star$ indicates terms not relevant for the moment. Set

$$\Theta(\vec{x}, \vec{x}') = (x^1, x^2, x^3 - a^3, \ldots, x^{\bar{m}} - a^{\bar{m}}, x_{1'}, \ldots, x_{\bar{m}'}).$$

Since Θ arises from a transformation on M that preserves ∇, Θ preserves $g_{\nabla,1}$. Since Θ preserves Ω, Θ also preserves J_+. Consequently, we may assume henceforth that

$$\tilde{P} = (a_1, a_2, 0, \ldots, 0, b_{1'}, b_{2'}, \ldots, b_{\bar{m}'}).$$

If $\Xi = (\Xi_{ij})$ is a linear transformation of $\mathbb{R}^{\bar{m}}$, let $\tilde{\Xi} = \tilde{\Xi}_{ij}$ be the inverse linear transformation so that $\Xi_{ij}\tilde{\Xi}_{ki} = \delta_j^k$. The induced transformation $\Theta_\Xi = \Xi \oplus \tilde{\Xi}$ of $(\vec{x}, \vec{x}')$ is given by:

$$\Theta_\Xi^* x^i = \Xi_{ij}x^j \quad \text{and} \quad \Theta_\Xi^* x_{i'} = \tilde{\Xi}_{ki}x_{k'} \text{ so}$$
$$\Theta_\Xi^*(dx^i \circ dx_{i'}) = \Xi_{ij}\tilde{\Xi}_{ki}dx^j \circ dx_{k'} = \delta_j^k dx^j \circ dx_{k'} \text{ and}$$
$$\Theta_\Xi^*(x_{i'}x_{j'}dx^i \circ dx^j) = \Xi_{ik}\Xi_{j\ell}\tilde{\Xi}_{ui}\tilde{\Xi}_{vj}x_{u'}x_{v'}dx^k \circ dx^\ell$$
$$= \delta_k^u \delta_\ell^v x_{u'}x_{v'}dx^k \circ dx^\ell = x_{k'}x_{\ell'}dx^k \circ dx^\ell.$$

Let $\Xi = \mathrm{Id}_2 \oplus \Xi_{\bar{m}-2}$ where $\Xi_{\bar{m}-2} \in \mathrm{GL}(\mathbb{R}^{\bar{m}-2})$ is a linear map of $\mathbb{R}^{\bar{m}-2}$. Since $\Xi_{\bar{m}-2}$ preserves $^0\nabla$, Ξ preserves $\nabla = \tilde{\nabla} \oplus {}^0\nabla$, so Θ preserves $g_{\nabla,1}$. Since Θ preserves Ω, Θ also preserves the almost para-complex structure J_+. The maps we shall consider subsequently are of this form and hence, are para-complex isometries. Let $f := (b_{3'}, \ldots, b_{\bar{m}'})$. Choose $\Xi_{\bar{m}-2} \in \mathrm{GL}(\mathbb{R}^{\bar{m}-2})$ so $\Xi_{\bar{m}-2} f = (c, 0, \ldots, 0)$ for some c. Let $\Xi = \mathrm{Id}_2 \oplus \Xi_{\bar{m}-2}$ and let $\Theta = \Theta_\Xi$. Assertion (1) now follows since

$$\Theta_{\mathrm{Id}_2 \oplus \Xi_{\bar{m}-2}} \tilde{P} = (a_1, a_2, 0, \ldots, 0, b_{1'}, b_{2'}, c, 0, \ldots, 0).$$

To simplify the notation, we replace c by $b_{3'}$. To prove Assertion (2), we assume $\bar{m} \geq 5$ and that $\tilde{P} = (a_1, a_2, 0, \ldots, 0, b_{1'}, b_{2'}, b_{3'}, 0, \ldots, 0)$. Let $\xi = (c_1, \ldots, c_{\bar{m}}, \star)$. Let e be $(c_4, c_5, \ldots, c_{\bar{m}})$ and choose $\Xi_{\bar{m}-3} \in \mathrm{GL}(\mathbb{R}^{\bar{m}-3})$ so that $\Xi_{\bar{m}-3} e = (c, 0, \ldots, 0)$ for some c. Let $\Xi = \mathrm{Id}_3 \oplus \Xi_{\bar{m}-3}$ and let $\Theta = \Theta_\Xi$. Then

$$\Theta \tilde{P} = \tilde{P} \text{ and } \Theta_* \xi = (c_1, c_2, c_3, c, 0, \ldots, 0, d_{1'}, d_{2'}, d_{3'}, d_{4'}, \ldots, d_{\bar{m}'}).$$

To simplify the notation, we shall replace c by c_4. Let $f = (d_{4'}, \ldots, d_{\bar{m}'}) = 0$. We may choose $\tilde{\Xi}_{\bar{m}-4} \in \mathrm{GL}(\mathbb{R}^{\bar{m}-4})$ so $\Xi_{\bar{m}-4} f = (c, 0, \ldots, 0)$ for some c. Let $\Xi = \mathrm{Id}_4 \oplus \Xi_{\bar{m}-4}$ and let $\Theta = \Theta_\Xi$. Assertion (2) now follows since

$$\Theta \tilde{P} = \tilde{P} \text{ and } \Theta_* \xi = (c_1, c_2, c_3, c_4, 0, \ldots, 0, d_{1'}, d_{2'}, d_{3'}, d_{4'}, c, 0, \ldots, 0).$$

Again, we replace c by $d_{5'}$ to simplify the notation. To prove Assertion (3), we assume $\bar{m} \geq 6$. Let $e_i := \partial_{x_i}$ and $f_i := \partial_{x_{i'}}$. Expand

$$\mathcal{J} e_i = \alpha_i^j e_j + \beta_i^j f_j \text{ and } \mathcal{J} f_i = \gamma_i^j e_j + \varrho_i^j f_j.$$

Fix an index $a \geq 6$ and let $\lambda \in \mathbb{R}$. Let $\Theta_{a,\lambda} = \Theta_{\mathrm{Id}_5 \oplus \Xi_{\bar{m}-5}(a,\lambda)}$ be given by:

$$\Theta_{a,\lambda}(e_i) = \left\{ \begin{array}{ll} \lambda e_i & \text{if } i = a \\ e_i & \text{if } i \neq a \end{array} \right\} \text{ and } \Theta(f_i) = \left\{ \begin{array}{ll} \lambda^{-1} f_i & \text{if } i = a \\ f_i & \text{if } i \neq a \end{array} \right\}.$$

Then $\Theta_{a,\lambda}$ is an isometry of $(T^*M, g_{\nabla,1})$, which fixes $(\tilde{P}, \xi)$. Consequently, $\Theta_{a,\lambda}$ commutes with $\mathcal{J}$ so

$$\alpha_i^a = \beta_i^a = \gamma_i^a = \varrho_i^a = \alpha_a^i = \beta_a^i = \gamma_a^i = \varrho_a^i = 0 \text{ for } i \neq a,$$
$$\beta_a^a = \gamma_a^a = 0.$$

If $\bar{m} \geq 7$, fix $5 < a < b$ and let $\Theta_{a,b} = \Theta_{\mathrm{Id}_5 \oplus \Xi_{\bar{m}-5}(a,b)}$ be given by

$$\Theta_{a,b}(e_i) = \left\{ \begin{array}{ll} e_b & \text{if } i = a \\ e_a & \text{if } i = b \\ e_i & \text{if } i \neq a \end{array} \right\} \text{ and } \Theta(f_i) = \left\{ \begin{array}{ll} f_b & \text{if } i = a \\ f_a & \text{if } i = b \\ f_i & \text{if } i \neq a \end{array} \right\}.$$

Again, $\Theta_{a,b}$ commutes with $\mathcal{J}$ and thus, $\alpha_a^a = \alpha_b^b$ and $\varrho_a^a = \varrho_b^b$. This shows that $\mathcal{J}$ has the following form relative to the basis $\{e_1, f_1, e_2, f_2, \ldots, e_{\bar{m}}, f_{\bar{m}}\}$:

$$
\mathcal{J} = \begin{pmatrix} \mathcal{J}_{10} & 0 & 0 & 0 & \cdots \\ 0 & \alpha & 0 & 0 & \cdots \\ 0 & 0 & \beta & 0 & \cdots \\ \cdots & \cdots & \cdots & \cdots & \cdots \end{pmatrix}. \tag{3.4.b}
$$

Since 0 and 1 are eigenvalues of multiplicity 1 and, since $\frac{1}{4}$ has multiplicity $2\bar{m} - 2$, we conclude $\alpha = \beta = \frac{1}{4}$.   □

By Theorem 3.20, $(T^*M, g_{\nabla,1})$ is not Jordan–Osserman on the zero section. More generally, we have that:

Theorem 3.22   Let $(\Sigma, \tilde{\nabla})$ be an affine Osserman surface with non-vanishing curvature tensor. Let $(M, \nabla) := (\Sigma \times \mathbb{R}^{\bar{m}-2}, \tilde{\nabla} \oplus {}^0\nabla)$. Then $(T^*M, g_{\nabla,1})$ is an Osserman manifold of signature $(\bar{m}, \bar{m})$ whose Jacobi operators have eigenvalues $\{0, 1, \frac{1}{4}, \ldots, \frac{1}{4}\}$ on $S^+(T^*M)$. Furthermore, $(T^*M, g_{\nabla,1})$ is not Jordan–Osserman at any point $\tilde{P} \in T^*M$.

Proof. Let $(\tilde{P}, \xi)$ be normalized as in Lemma 3.21 (1,2). By Lemma 3.21 (3), (see Equation (3.4.b)), we have that:

$$
\mathcal{J}(\xi) := {}^{g_{\nabla,1}}\mathcal{J}(\xi)_{\tilde{P}} = \begin{pmatrix} \mathcal{J}_{10} & 0 \\ 0 & \frac{1}{4}\mathrm{Id}_{2\bar{m}-10} \end{pmatrix}.
$$

This reduces the study of the Jordan normal form to the case $\bar{m} = 5$ and $\mathcal{J} = \mathcal{J}_{10}$. Set

$$
\mathcal{A}(\xi) = \mathcal{J}(\xi) \cdot (\mathcal{J}(\xi) - \mathrm{Id}) \cdot \left(\mathcal{J}(\xi) - \tfrac{1}{4}\mathrm{Id}\right).
$$

We adopt the notation of Equation (3.4.a). We consider the unit spacelike vectors:

$$
\tilde{\xi} = \varepsilon^{-1/2}(0, 0, 0, \varepsilon, 0, 0, 0, 0, \tfrac{1}{2}, 0),
$$
$$
\hat{\xi} = \varepsilon^{-1/2}(1, 0, 0, 1, 0, \tfrac{1}{2}(\varepsilon - b_1^2 - 2b_1 \partial_{x_1}\theta(a_1, a_2)), 0, 0, 0, 0).
$$

Let $(\mathcal{A}(\hat{\xi}))_{42}$ be the $(4, 2)$-entry of the $(\mathcal{A}(\hat{\xi}))$-matrix. One computes that

$$
\mathcal{A}(\tilde{\xi}) = 0, \qquad (\mathcal{A}(\hat{\xi}))_{42} = -\varepsilon^{-1}\tfrac{3}{16}\partial_{x_1}\partial_{x_2}\theta(a_1, a_2) \neq 0.
$$

Thus, $\mathcal{J}(\tilde{\xi})$ is diagonalizable while $\mathcal{J}(\hat{\xi})$ is not diagonalizable. This shows that $(T^*M, g_{\nabla,1})$ is not Jordan–Osserman at any point $\tilde{P} \in T^*M$.   □

FURTHER ANALYSIS OF THE JORDAN NORMAL FORM

We conclude this section by giving a complete analysis of the Jordan normal form for the manifold of Theorem 3.22. There is no restriction in assuming $\tilde{m} = 5$ and taking $(\tilde{P}, \xi)$ normalized as in Lemma 3.21. As in the previous theorem, set

$$\mathcal{A}^r(\xi) = \mathcal{J}(\xi) \cdot (\mathcal{J}(\xi) - g_{\nabla,1}(\xi, \xi) \, \mathrm{Id}) \cdot \left(\mathcal{J}(\xi) - \tfrac{1}{4} g_{\nabla,1}(\xi, \xi) \, \mathrm{Id} \right)^r, \quad r \geq 1.$$

We now examine the possible values of $\mathcal{A}^r(\xi)$ for $r = 1, 2, 3$. From now on, we assume that all objects defined on Σ (θ and its partial derivatives) are evaluated at (a_1, a_2) and put $\varepsilon_\xi = g_{\nabla,1}(\xi, \xi)$. Then

$$\mathcal{A}(\xi) = \frac{3}{16} \varepsilon_\xi \partial_{x_1} \partial_{x_2} \theta \left(\begin{array}{c|c} S & 0 \\ \hline T & {}^t S \end{array} \right),$$

where the matrix S is given by

$$S = \begin{pmatrix} -c_1 c_2 (\nu_1 + \varepsilon_\xi) & c_1^2 (\nu_1 + \varepsilon_\xi) & 0 & 0 & 0 \\ -c_2^2 (\nu_1 + \varepsilon_\xi) & c_1 c_2 (\nu_1 + \varepsilon_\xi) & 0 & 0 & 0 \\ -c_2 c_3 \nu_1 & c_1 c_3 \nu_1 & 0 & 0 & 0 \\ -c_2 c_4 \nu_1 & c_1 c_4 \nu_1 & 0 & 0 & 0 \\ 0 & 0 & 0 & 0 & 0 \end{pmatrix},$$

with $\nu_1 = b_1 b_3 c_1 c_3 + b_2 b_3 c_2 c_3 + b_3^2 c_3^2 + 2 c_3 d_3 + 2 c_4 d_4 - \varepsilon_\xi$, and the matrix T is given by:

$$\begin{pmatrix} \dfrac{-c_2 t_{11}}{3 \varepsilon_\xi \partial_{x_1} \partial_{x_2} \theta} & \dfrac{t_{12}}{3 \varepsilon_\xi \partial_{x_1} \partial_{x_2} \theta} & c_2 (d_3 \nu_1 + (b_3^2 c_3 + d_3 - \nu_2) \varepsilon_\xi) & c_2 d_4 \nu_1 & c_2 d_5 \nu_1 \\ \dfrac{t_{21}}{3 \varepsilon_\xi \partial_{x_1} \partial_{x_2} \theta} & \dfrac{-c_1 t_{22}}{3 \varepsilon_\xi \partial_{x_1} \partial_{x_2} \theta} & -c_1 (d_3 \nu_1 + (b_3^2 c_3 + d_3 - \nu_2) \varepsilon_\xi) & -c_1 d_4 \nu_1 & -c_1 d_5 \nu_1 \\ c_2 \nu_1 \nu_2 & -c_1 \nu_1 \nu_2 & 0 & 0 & 0 \\ c_2 d_4 \nu_1 & -c_1 d_4 \nu_1 & 0 & 0 & 0 \\ c_2 d_5 \nu_1 & -c_1 d_5 \nu_1 & 0 & 0 & 0 \end{pmatrix},$$

where

$$\nu_2 = b_1 b_3 c_1 + b_2 b_3 c_2 + b_3^2 c_3 + d_3,$$

$$\nu = \partial_{x_1} \partial_{x_2} \theta (3 b_1 \partial_{x_2} \theta + 3 b_2 \partial_{x_1} \theta + 4 \partial_{x_1} \partial_{x_2} \theta) - 3 b_1 \partial_{x_1} \partial_{x_2} \partial_{x_2} \theta + 3 b_2 \partial_{x_1} \partial_{x_1} \partial_{x_2} \theta,$$

$$t_{11} = c_2 \nu + 4 c_2 (\partial_{x_1} \partial_{x_2} \theta)^2 \{ 1 + 3 \nu_1^2 + 2 \varepsilon_\xi (b_3^2 c_3^2 - 2 c_4 d_4 + 2 \nu_1 - 2 c_3 \nu_2) \}$$
$$- 3 \partial_{x_1} \partial_{x_2} \theta \{ b_1 (b_1 c_1 + b_2 c_2 + 2 b_3 c_3) + \varepsilon_\xi \nu_1 (b_1 (b_1 c_1 + b_2 c_2 + b_3 c_3) + 2 d_1) $$
$$+ 2 d_1 + 2 b_1 c_1 (1 + \varepsilon_\xi \nu_1) \partial_{x_1} \theta \},$$

$$t_{22} = c_1 \nu + 4 c_1 (\partial_{x_1} \partial_{x_2} \theta)^2 \{ 1 + 3 \nu_1^2 + 2 \varepsilon_\xi (b_3^2 c_3^2 - 2 c_4 d_4 + 2 \nu_1 - 2 c_3 \nu_2) \}$$
$$+ 3 \partial_{x_1} \partial_{x_2} \theta \{ b_2 (b_1 c_1 + b_2 c_2 + 2 b_3 c_3) + \varepsilon_\xi \nu_1 (b_2 (b_1 c_1 + b_2 c_2 + b_3 c_3) + 2 d_2) $$
$$+ 2 d_2 - 2 b_2 c_2 (1 + \varepsilon_\xi \nu_1) \partial_{x_2} \theta \},$$

$$t_{12} = c_1 c_2 \nu + 4 c_1 c_2 (\partial_{x_1} \partial_{x_2} \theta)^2 \{1 + 3\nu_1^2 + 2\varepsilon_\xi (b_3^2 c_3^2 - 2c_4 d_4 + 2\nu_1 - 2c_3 \nu_2))\}$$
$$+ 3\partial_{x_1} \partial_{x_2} \theta \{b_1 b_2 c_1 c_2 + \varepsilon_\xi \nu_1 (b_1 b_2 c_1 c_2 + c_1 d_1 + c_2 (b_2^2 c_2 + b_2 b_3 c_3 + 3d_2) + \nu_1)$$
$$+ c_1 d_1 + c_2 (b_2^2 c_2 + 2b_2 b_3 c_3 + 3d_2) + \nu_1 - 2b_2 c_2^2 (1 + \varepsilon_\xi \nu_1) \partial_{x_2} \theta\},$$
$$t_{21} = t_{12} + 3\partial_{x_1} \partial_{x_2} \theta \{b_3^2 c_3^2 - 2(c_1 d_1 + c_2 d_2 + c_4 d_4 + c_3 \nu_2)$$
$$- \varepsilon_\xi (\nu_1 (2c_1 d_1 + 2c_2 d_2 + \nu_1) - 1)\}.$$

Hence,

$$\mathcal{A}^2(\xi) = \tfrac{3}{8} \varepsilon_\xi (\partial_{x_1} \partial_{x_2} \theta)^2 \, \psi
\left(
\begin{array}{ccccc|c}
 & & 0 & & & 0 \\
\hline
c_2^2 & -c_1 c_2 & 0 & 0 & 0 & \\
-c_1 c_2 & c_1^2 & 0 & 0 & 0 & \\
0 & 0 & 0 & 0 & 0 & 0 \\
0 & 0 & 0 & 0 & 0 & \\
0 & 0 & 0 & 0 & 0 &
\end{array}
\right),$$

where

$$\psi = b_3^2 c_3^2 \{c_1 d_1 + c_2 d_2 - c_3 d_3 - c_4 d_4 + b_1 c_1^2 \partial_{x_1} \theta - b_2 c_2^2 \partial_{x_2} \theta\}$$
$$- (2c_3 d_3 + 2c_4 d_4 - \varepsilon_\xi) \{b_3 c_3 (b_1 c_1 + b_2 c_2) + (c_3 d_3 + c_4 d_4)\}.$$

Thus $\mathcal{A}^3(\xi) = 0$. This shows that Rank$\{\mathcal{A}(\xi)^2\} \leq 1$, and Rank$\{\mathcal{A}(\xi)\} \leq 2$. Therefore, the Jacobi operator corresponding to a given vector ξ is either diagonalizable, it contains a single 2×2 or 3×3 Jordan block, or it contains two 2×2 Jordan blocks, depending on the vector and the basepoint. All these possibilities, except that of a single 2×2 Jordan block, can be realized at any point. Let

$$\tilde{P} = (a_1, a_2, 0, 0, 0, b_1, b_2, b_3, 0, 0) \text{ and } \xi = (c_1, c_2, c_3, c_4, 0, d_1, d_2, d_3, d_4, d_5).$$

1. If $\psi \neq 0$, and $(c_1, c_2) \neq (0, 0)$, then the Jacobi operator has a 3×3 Jordan block. For example, take: $\xi = (1, 0, 0, 1, 0, \tfrac{1}{2}(\varepsilon - b_1^2 - 2b_1 \partial_{x_1} \theta - 2), 0, 0, 1, 0)$, $g_{\nabla,1}(\xi, \xi) = \varepsilon$.

2. If $c_1 = c_2 = 0$, then the Jacobi operators are diagonalizable. For example, we could take $\xi = (0, 0, 0, \varepsilon, 0, 0, 0, 0, \tfrac{1}{2}, 0)$, $g_{\nabla,1}(\xi, \xi) = \varepsilon$.

3. If $\psi = 0$, and $c_1 = c_2 = 0$ does not hold, then the Jacobi operator corresponding to the vectors $\xi = (c_1, c_2, 0, 0, 0, d_1, d_2, -b_3(b_1 c_1 + b_2 c_2), 0, 0)$ is either diagonalizable or it has a 2×2 Jordan block, depending on whether ν vanishes or not where

$$\nu = \partial_{x_1} \partial_{x_2} \theta (3b_1 \partial_{x_2} \theta + 3b_2 \partial_{x_1} \theta + 4\partial_{x_1} \partial_{x_2} \theta) - 3b_1 \partial_{x_1} \partial_{x_2} \partial_{x_2} \theta + 3b_2 \partial_{x_1} \partial_{x_1} \partial_{x_2} \theta.$$

The Jacobi operator corresponding to any other unit vector has two 2×2 Jordan blocks. For example, $\xi = (1, 0, 0, 1, 0, \tfrac{1}{2}(\varepsilon - b_1^2 - 2b_1 \partial_{x_1} \theta), 0, 0, 0, 0)$, $g_{\nabla,1}(\xi, \xi) = \varepsilon$.

4. Finally, at any point where $\nu = 0$, the Jacobi operators never have a unique 2×2 Jordan block.

3.5   (SEMI) PARA-COMPLEX OSSERMAN MANIFOLDS

If (M, g, J_+) is an almost para-complex manifold, let $\mathcal{H} \subset \mathrm{Gr}_2(TM)$ be the Grassmannian of all para-complex 2-planes. If $\pi \in \mathrm{Gr}_2(TM)$, then $\pi \in \mathcal{H}$ if and only if $J_+\pi = \pi$. If e_1 is any unit spacelike vector, then $\pi := \mathrm{Span}\{e_1, J_+e_1\}$ belongs to $\mathcal{H}$ and all elements of $\mathcal{H}$ arise in this way. We adopt the notation of Equation (1.11.a) and define the *para-complex Jacobi operator* by setting:

$$\mathcal{J}(\pi) := \mathcal{J}(e_1) - \mathcal{J}(J_+e_1).$$

One says that (M, g, J_+) is *semi para-complex Osserman* if $\mathcal{J}$ has constant eigenvalues on $\mathcal{H}$. If, in addition, $\mathcal{J}(\pi)$ commutes with the almost para-complex structure J_+, then (M, g, J_+) is said to be *para-complex Osserman*. There are similar notions in the complex setting, which we shall not pursue further; we refer to Brozos-Vázquez, García-Río, and Gilkey [29] for further details. The modified Riemannian extensions provide examples of semi para-complex manifolds in the neutral signature context, which are para-complex if and only if the underlying affine connection is flat.

Theorem 3.23   *Let (M, ∇) be an affine manifold of dimension $\bar{m}$. Give the associated modified Riemannian extension $(T^*M, g_{\nabla,1})$ the almost para-complex structure J_+ of Equation (3.2.b). Let $\mathcal{J}(\pi)$ be the associated higher order Jacobi operator for $\pi \in \mathcal{H}(T^*M, g_{\nabla,1})$.*

1. *The eigenvalues of $\mathcal{J}(\pi)$ are $(1, \frac{1}{2})$ with multiplicities $(2, 2\bar{m} - 2)$ respectively.*

2. *The Jordan blocks of $\mathcal{J}(\pi)$ have size at most 2×2.*

3. *$(T^*M, g_{\nabla,1}, J_+)$ is semi para-complex Osserman.*

4. *$\mathcal{J}(\pi)J_+ = J_+\mathcal{J}(\pi)$ for all π if and only if ∇ is flat.*

5. *Let $\bar{m} \geq 3$. There exists an affine Osserman manifold (M, ∇) of dimension $\bar{m}$ so that $(T^*M, g_{\nabla,1}, J_+)$ is not Jordan semi para-complex Osserman, and so that the para-complex Jacobi operators are not always diagonalizable.*

Proof. Let (M, ∇) be an affine manifold. Let $\tilde{Q} \in T^*M$, let $\xi \in S^+(T_{\tilde{Q}}T^*M, g_{\nabla,1})$, and let $\pi = \mathrm{Span}\{\xi, J_+\xi\}$. Let $a = \sigma_*\xi = \sigma_*J_+(\xi)$. By Lemma 3.16,

$$^2\mathcal{J}(\xi) = \begin{pmatrix} ^\nabla\mathcal{J}(a) & 0 \\ \star & {}^t({}^\nabla\mathcal{J}(a)) \end{pmatrix}, \qquad ^2\mathcal{J}(J_+\xi) = \begin{pmatrix} ^\nabla\mathcal{J}(a) & 0 \\ \star_1 & {}^t({}^\nabla\mathcal{J}(a)) \end{pmatrix},$$

$$^2\mathcal{J}(\pi) = {}^2\mathcal{J}(\xi) - {}^2\mathcal{J}(J_+\xi) = \begin{pmatrix} 0 & 0 \\ \star_\pi & 0 \end{pmatrix}.$$

Let $\mathfrak{D} := \ker\{\sigma_*\} = \mathrm{Span}\{\partial_{x^{1'}}, \ldots, \partial_{x^{\tilde{m}'}}\}$. Then $\mathrm{Range}\{{}^2\mathcal{J}(\pi)\} \subset \mathfrak{D}$ and ${}^2\mathcal{J}(\pi)\mathfrak{D} = 0$. This shows that ${}^2\mathcal{J}(\pi)$ is nilpotent. Choose a basis

$$\{e_1, e_2, f_1, \ldots, f_{2\tilde{m}-2}\} \text{ for } T_{\tilde{Q}}T^*M$$

so $\mathrm{Span}\{e_1, e_2\}$ is the $+1$ eigenspace of ${}^{g_0}\mathcal{J}(\pi)$ and so $\mathrm{Span}\{f_1, \ldots, f_{2\tilde{m}-2}\}$ is the $+\frac{1}{2}$ eigenspace of ${}^{g_0}\mathcal{J}(\pi)$. We compute:

$$({}^{g_{\nabla,1}}\mathcal{J}(\pi) - \mathrm{Id})e_i = {}^2\mathcal{J}(\pi)e_i \in \mathfrak{D} \quad \text{and} \quad ({}^{g_{\nabla,1}}\mathcal{J}(\pi) - \tfrac{1}{2}\,\mathrm{Id})f_i = {}^2\mathcal{J}(\pi)f_i \in \mathfrak{D}.$$

By Equation (3.3.a),

$$ {}^{g_0}\mathcal{J}(\pi) = \mathrm{Id} \text{ on } (\xi - J_+\xi) \cdot \mathbb{R} \quad \text{and} \quad {}^{g_0}\mathcal{J}(\pi) = \tfrac{1}{2}\,\mathrm{Id} \text{ on } \mathcal{S}(\xi). $$

By Lemma 3.17, ${}^{g_0}\mathcal{J}(\pi)\mathfrak{D} \subset \mathfrak{D}$. As ${}^2\mathcal{J}(\pi) = 0$ on $\mathfrak{D}$, ${}^{g_{\nabla,1}}\mathcal{J}(\pi)\mathfrak{D} \subset \mathfrak{D}$ and:

$$\mathrm{Range}\{({}^{g_{\nabla,1}}\mathcal{J}(\pi) - \mathrm{Id}) \cdot ({}^{g_{\nabla,1}}\mathcal{J}(\pi) - \tfrac{1}{2}\,\mathrm{Id})\} \subset \mathfrak{D}.$$

Since ${}^2\mathcal{J}(\pi) = 0$ on $\mathfrak{D}$, ${}^{g_{\nabla,1}}\mathcal{J}(\pi) = {}^{g_0}\mathcal{J}(\pi)$ on $\mathfrak{D}$. Equation (3.3.a) and Lemma 3.17 yield:

$$({}^{g_{\nabla,1}}\mathcal{J}(\pi) - \mathrm{Id}) \cdot ({}^{g_{\nabla,1}}\mathcal{J}(\pi) - \tfrac{1}{2}\,\mathrm{Id})\mathfrak{D} = \{0\} \quad \text{so}$$
$$({}^{g_{\nabla,1}}\mathcal{J}(\pi) - \mathrm{Id})^2 \cdot ({}^{g_{\nabla,1}}\mathcal{J}(\pi) - \tfrac{1}{2}\,\mathrm{Id})^2 = \{0\}.$$

Consequently $\mathrm{Spec}\{{}^{g_{\nabla,1}}\mathcal{J}(\pi)\} \subset \{\frac{1}{2}, 1\}$ and ${}^{g_{\nabla,1}}\mathcal{J}(\pi)$ has only 1×1 or 2×2 Jordan blocks. As in the proof of Theorem 3.15, we set ${}^\varepsilon\nabla := \varepsilon\nabla + (1 - \varepsilon)^0\nabla$ to construct a 1-parameter family of semi para-complex Osserman metrics $g_{\varepsilon\nabla}$ interpolating between $g_{\nabla,1}$ and g_0. As the eigenvalues are unchanged, the eigenvalue multiplicities are unchanged. Consequently, $\frac{1}{2}$ is an eigenvalue of multiplicity $2\tilde{m} - 2$ and 1 is an eigenvalue of multiplicity 2. Assertions (1)-(3) now follow.

Generalizing the para-Kähler identity discussed in Section 1.4, one says that an almost para-Hermitian manifold satisfies the *third Gray identity* if the curvature tensor is invariant under the almost para-complex structure:

$$R(J_+ \cdot, J_+ \cdot, J_+ \cdot, J_+ \cdot) = R(\cdot, \cdot, \cdot, \cdot).$$

The third Gray identity in the complex setting is crucial (see, for example, the discussion in Brozos-Vázquez, García-Río, and Gilkey [29], in Di Scala and Vezzoni [72], and in Di Scala, Lauret, and Vezzoni [73]). A purely algebraic computation shows that an almost para-Hermitian manifold (in our case $(T^*M, g_{\nabla,1}, J_+)$) satisfies the third Gray identity if and only if we have the commutation relation $J_+\mathcal{J}(\pi) = \mathcal{J}(\pi)J_+$ for all $\pi \in \mathcal{H}(T^*M, J_+)$.

Suppose that $(T^*M, g_{\nabla,1}, J_+)$ satisfies the third Gray identity. Fix a point $\tilde{Q}$ in the zero section of T^*M with $P = \sigma(\tilde{Q})$. Use Lemma 1.6 to choose coordinates on M so ${}^\nabla\Gamma(P) = 0$. We apply Lemma 3.19. Since $\tilde{\mathbb{C}}P$ satisfies the third Gray identity, we conclude ${}^{g_\nabla}R$ satisfies the third

Gray identity at $\tilde{Q}$. Thus

$$
\begin{aligned}
{}^{g\nabla}R(\partial_{x_i}, \partial_{x_j}, \partial_{x_k}, \partial_{x^{\ell'}})(\tilde{Q}) &= {}^{g\nabla}R(J_+\partial_{x_i}, J_+\partial_{x_j}, J_+\partial_{x_k}, J_+\partial_{x^{\ell'}})(\tilde{Q}) \\
&= {}^{g\nabla}R(\partial_{x_i}, \partial_{x_j}, \partial_{x^{\ell'}}, \partial_{x_k})(\tilde{Q}) \\
&= {}^{g\nabla}R(\partial_{x_i}, \partial_{x_j}, \partial_{x^{\ell'}}, \partial_{x_k})(\tilde{Q}) .
\end{aligned}
$$

Consequently ${}^{g\nabla}R(\tilde{Q}) = 0$. By Lemma 3.19, this implies $\nabla\mathcal{R}(P) = 0$ and since P was arbitrary, (M, ∇) is flat, which completes the proof; the converse implication is trivial. Assertion (4) now follows.

We now establish Assertion (5). Let $\bar{m} \geq 3$, let $M = \mathbb{R}^{\bar{m}}$, let $P = 0$, and let $\tilde{Q} = (0, 0)$. Let $\theta = \theta(x^1)$ be a smooth function of a single variable. Suppose $\theta(0) = 0$ and $\partial_{x_1}\theta(0) \neq 0$. Define an affine connection ∇ by requiring that the only non-zero Christoffel symbol is $\nabla\Gamma_{22}{}^3 = \theta$. Since $\theta = \theta(x^1)$ and the only non-zero covariant derivative is $\nabla_{\partial_{x_2}}\partial_{x_2} = \theta\,\partial_{x_3}$, the only non-zero curvature is $\nabla\mathcal{R}(\partial_{x_1}, \partial_{x_2})\partial_{x_2} = (\partial_{x_1}\theta)\partial_{x_3}$. Hence, the associated Jacobi operators $\nabla\mathcal{J}$ are nilpotent and (M, ∇) is affine Osserman. By Lemma 3.19, the non-zero components of the curvature tensor at $\tilde{Q}$ (which is on the zero section of T^*M) are:

$$
\begin{aligned}
{}^{g\nabla,1}R_{ii'i'i}(\tilde{Q}) &= -1, & {}^{g\nabla,1}R_{3'221}(\tilde{Q}) &= \partial_{x_1}\theta, \\
{}^{g\nabla,1}R_{ij'i'j}(\tilde{Q}) &= {}^{g\nabla,1}R_{ii'j'j}(\tilde{Q}) = -\tfrac{1}{2} & (i \neq j).
\end{aligned}
$$

We first consider

$$
\xi := \tfrac{1}{\sqrt{2}}(\partial_{x_1} + \partial_{x^{1'}}), \quad J_+\xi = \tfrac{1}{\sqrt{2}}(\partial_{x_1} - \partial_{x^{1'}}), \quad \pi_\xi := \mathrm{Span}\{\xi, J_+\xi\}.
$$

As ${}^2\mathcal{J}(\xi) = {}^2\mathcal{J}(J_+\xi) = 0, \mathcal{J}(\xi) = {}^{g0}\mathcal{J}(\xi)$ and $\mathcal{J}(\pi_\xi) = {}^{g0}\mathcal{J}(\pi_\xi)$ are diagonalizable. Next consider

$$
\eta := \tfrac{1}{2}(\partial_{x_1} + \partial_{x_3} + \partial_{x^{1'}} + \partial_{x^{3'}}), \quad J_+\eta = \tfrac{1}{2}(\partial_{x_1} + \partial_{x_3} - \partial_{x^{1'}} - \partial_{x^{3'}}), \quad \pi_\eta := \mathrm{Span}\{\xi, J_+\xi\}.
$$

The only non-trivial components of the Jacobi operators are $\mathcal{J}(\eta)\partial_{x_2} = \tfrac{1}{2}\partial_{x_1}\theta\,\partial_{x^{2'}}$ and $\mathcal{J}(J_+\eta)\partial_{x_2} = -\tfrac{1}{2}\partial_{x_1}\theta\,\partial_{x^{2'}}$. Consequently:

$$
\mathcal{J}(\pi_\eta)\partial_{x_2} = (\partial_{x_1}\theta)\partial_{x^{2'}} .
$$

Since $\pi_2 := \mathrm{Span}\{\partial_{x_2}, \partial_{x^{2'}}\}$ is contained both in the $\tfrac{1}{4}$-eigenspace of ${}^{g0}\mathcal{J}(\eta)$ and in the $\tfrac{1}{2}$-eigenspace of ${}^{g0}\mathcal{J}(\pi_\eta)$, we see that both $\mathcal{J}(\eta)$ and $\mathcal{J}(\pi_\eta)$ exhibit non-trivial Jordan normal form. $\square$

CHAPTER 4

(para)-Kähler–Weyl Manifolds

In Chapter 4 we report on work of Gilkey and Nikčević [102, 103, 104, 105] and of Brozos-Vázquez *et al.* [32]. We work in the complex and in the para-complex contexts to examine (para)-Kähler–Weyl structures. If the underlying dimension is at least six, then any (para)-Kähler–Weyl algebraic curvature tensor is in fact Riemannian. Consequently, any (para)-Kähler–Weyl structure is trivial in the geometric setting as well. The 4-dimensional setting is quite different. Since every (para)-Kähler–Weyl algebraic curvature tensor is geometrically realizable, and since every 4-dimensional (para)-Hermitian manifold admits a unique (para)-Kähler–Weyl structure, these structures can be non-trivial. We shall always work in both the complex and the para-complex settings and shall, for the most part, attempt to treat these two cases in parallel.

4.1 NOTATIONAL CONVENTIONS

We adopt the following notational conventions. Throughout Chapter 4, $(M, g, J_\pm)$ will be a (para)-Hermitian manifold. This means that (M, g) is a pseudo-Riemannian manifold, that $J_\pm$ is an integrable (para)-complex structure on M, and that $J_\pm^* g = \mp g$. The (para)-Kähler form $\Omega_\pm = \Omega_\pm(M, g, J_\pm)$ is characterized by the identity:

$$\Omega_\pm(X, Y) = g(X, J_\pm Y) \quad \text{for } X, Y \in C^\infty(TM);$$

it plays a central role. We shall assume $H^1_{\text{DeR}}(M) = 0$ henceforth, to avoid difficulties with the fundamental group; consequently, if ϕ is a smooth 1-form with $d\phi = 0$, then we can find f so $df = \phi$. Let ∇ be a Weyl connection; this means that ∇ is an affine connection so that

$$\nabla g = -2\phi \otimes g \text{ where } \phi \in C^\infty(T^*M) \text{ is the } associated \ 1\text{-}form.$$

We say that the quadruple $(M, g, J_\pm, \nabla)$ is a *(para)-Kähler–Weyl manifold* if $\nabla J_\pm = 0$. The structure is said to be a *trivial (para)-Kähler–Weyl structure* if there is a conformally equivalent metric g_1 so that $\nabla = {}^{g_1}\nabla$ is the associated Levi-Civita connection; $(M, g_1, J_\pm)$ is then a (para)-Kähler manifold. By Theorem 1.11, the Weyl structure is trivial if and only if $d\phi = 0$ or, equivalently, if the alternating Ricci tensor ρ_a vanishes.

We recall the following notational conventions which were established in Section 1.3 and in Section 1.4. Let $(V, \langle \cdot, \cdot \rangle)$ be an inner product space of dimension m. Let $\mathfrak{W} \subset \otimes^4 V^*$ be the space of Weyl algebraic curvature tensors; these are the tensors satisfying:

$$A(x, y, z, w) + A(y, x, z, w) = 0,$$
$$A(x, y, z, w) + A(y, z, x, w) + A(z, x, y, w) = 0,$$
$$A(x, y, z, w) + A(x, y, w, z) = -\frac{4}{m}\rho_a(x, y)\langle z, w\rangle.$$

Let $\mathfrak{R} \subset \mathfrak{W}$ be the subspace of Riemannian algebraic curvature tensors satisfying additionally:

$$A(x, y, z, w) + A(x, y, w, z) = 0.$$

Let $\mathfrak{K}_\pm \subset \otimes^4 V^*$ denote the space of tensors satisfying the (para)-Kähler identity:

$$A(x, y, z, w) = \mp A(x, y, J_\pm z, J_\pm w). \tag{4.1.a}$$

The spaces of (para)-Kähler Riemannian tensors $\mathfrak{K}_{\pm,\mathfrak{R}}$ and of (para)-Kähler–Weyl tensors $\mathfrak{K}_{\pm,\mathfrak{W}}$ are obtained by imposing the (para)-Kähler identity on $\mathfrak{R}$ and on $\mathfrak{W}$, respectively:

$$\mathfrak{K}_{\pm,\mathfrak{W}} = \mathfrak{W} \cap \mathfrak{K}_\pm \text{ and } \mathfrak{K}_{\pm,\mathfrak{R}} = \mathfrak{R} \cap \mathfrak{K}_\pm.$$

If the dimension is at least six, then no new phenomena arise. Section 4.2 is devoted to the proof of the following result; Assertion (2) in the Riemannian setting is originally due to Pedersen, Poon, and Swann [178] and to Vaisman [201, 202]; they used different approaches to the subject than we shall employ:

Theorem 4.1

1. If $(V, \langle\cdot, \cdot\rangle, J_\pm, A)$ is a (para)-Kähler–Weyl curvature model with $\dim(V) \geq 6$, then $A \in \mathfrak{R}$. Consequently $\mathfrak{K}_{\pm,\mathfrak{W}} = \mathfrak{K}_{\pm,\mathfrak{R}}$.

2. If $(M, g, J_\pm, \nabla)$ is a (para)-Kähler Weyl manifold with $\dim(M) \geq 6$ and with $H^1_{\mathrm{DeR}}(M) = 0$, then the associated Weyl structure is trivial.

Let δ be the coderivative defined in Equation (1.1.b); we expressed δ in terms of the Hodge $\star$ operator in Equation (1.8.b) setting $\delta = -\star d\star$ on 2-forms. We will establish the following result in Section 4.3; it is due to Kokarev and Kotschick [141] in the Riemannian setting; they used a quite different approach than we shall use:

Theorem 4.2   *Every 4-dimensional (para)-Hermitian manifold $(M, g, J_\pm)$ such that $H^1_{\mathrm{DeR}}(M) = 0$ admits a unique (para)-Kähler–Weyl structure with associated 1-form given by $\phi_\pm = \pm\frac{1}{2}J_\pm^*\delta\Omega_\pm$.*

By Theorem 1.24, we have a short exact sequence $0 \to \mathfrak{R} \to \mathfrak{W} \to \rho_a\{\mathfrak{W}\} \to 0$. This gives rise to the short exact sequence:

$$0 \to \mathfrak{K}_{\pm,\mathfrak{R}} \to \mathfrak{K}_{\pm,\mathfrak{W}} \to \rho_a\{\mathfrak{K}_{\pm,\mathfrak{W}}\} \to 0. \tag{4.1.b}$$

Thus, attention is focused on the alternating Ricci tensor, which takes values in Λ^2. We consider the structure groups which were defined in Section 1.5:

$$\mathcal{U}_\pm := \{T \in \mathcal{O} : T J_\pm = J_\pm T\},$$
$$\mathcal{U}_\pm^\star := \{T \in \mathcal{O} : T J_\pm = J_\pm T \text{ or } T J_\pm = -J_\pm T\}.$$

The group $\mathcal{U}_+$ is the *para-unitary group* and the group $\mathcal{U}_-$ is the *unitary group*. The groups $\mathcal{U}_\pm^\star$ are $\mathbb{Z}_2$ extensions of these groups; we let $\chi(T) \in \mathbb{Z}_2$ be defined by Equation (1.5.a):

$$T J_\pm = \chi(T) J_\pm T.$$

It is convenient to be able to interchange the structures $J_\pm$ and $-J_\pm$. We recall the following definitions from Equation (1.5.d):

$$\Lambda^2_{\mp,0} := \{\theta \in \Lambda^2 : \theta \perp \Omega_\pm, \ J_\pm^* \theta = \mp \theta\}, \text{ and } \Lambda^2_\pm := \{\theta \in \Lambda^2 : J_\pm^* \theta = \pm \theta\}.$$

By Lemma 1.17, we have an orthogonal direct sum decomposition of

$$\Lambda^2 = \Omega_\pm \cdot \mathbb{R} \oplus \Lambda^2_{\mp,0} \oplus \Lambda^2_\pm$$

as the direct sum of inequivalent and irreducible $\mathcal{U}_\pm^\star$ modules. We shall show in Section 4.2, whilst proving Theorem 4.2, that the module $\Omega_\pm \cdot \mathbb{R}$ plays no role and that

$$\rho_a\{\mathfrak{R}_\pm, \mathfrak{W}\} \subset \Lambda^2_\pm \oplus \Lambda^2_{\mp,0}. \tag{4.1.c}$$

The question of providing homogeneous examples was posed to us by Prof. Alekseevsky and we are grateful to him for the suggestion. We note that Calvaruso [40] undertook a related 3-dimensional problem by examining 3-dimensional homogeneous Lorentzian metrics with a prescribed Ricci tensor. We will follow the discussion in Brozos-Vázquez *et al.* [32] to establish the following result in Section 4.4:

Theorem 4.3

1. *Let $(V, \langle \cdot, \cdot \rangle, J_+)$ be a 4-dimensional para-Hermitian vector space of signature $(2, 2)$. Then every element of $\Lambda^2_{-,0} \oplus \Lambda^2_+$ is geometrically realizable as the alternating Ricci tensor of the para-Kähler–Weyl structure of a suitably chosen 4-dimensional Lie group with a left-invariant para-Hermitian structure of signature $(2, 2)$.*

2. *Let $(V, \langle \cdot, \cdot \rangle, J_-)$ be a 4-dimensional Hermitian vector space of signature $(0, 4)$. Then every element of $\Lambda^2_{+,0} \oplus \Lambda^2_-$ is geometrically realizable as the alternating Ricci tensor of the Kähler–Weyl structure of a suitable chosen 4-dimensional Lie group with a left invariant Hermitian structure of signature $(0, 4)$.*

We shall use Theorem 4.3 to prove the following result in Section 4.5; the proof originally given by Gilkey and Nikčević [103, 105] was quite different and did not employ the examples of Theorem 4.3.

Theorem 4.4   *Let* $(V, \langle \cdot, \cdot \rangle, J_\pm)$ *be a 4-dimensional (para)-Hermitian vector space of dimension four. Then* $\mathfrak{K}_{\pm, \mathfrak{W}} = \mathfrak{K}_{\pm, \mathfrak{R}} \oplus L_\pm^2$ *where* ρ_a *provides a* $\mathcal{U}_\pm^\star$ *module isomorphism from* $L_\pm^2$ *to* $\Lambda_{\mp, 0}^2 \oplus \Lambda_\pm^2$.

Theorem 4.4 is one of the facts about 4-dimensional geometry that distinguishes it from the higher-dimensional setting; the module $L_\pm^2$ provides additional curvature possibilities in dimension $m = 4$ and is the algebraic reason why Theorem 4.1 fails if $m = 4$.

All the algebraic possibilities of Theorem 4.4 can be realized geometrically. In Section 4.6, we will use Theorem 4.3 and Theorem 4.4 to establish the following result; again, the proof originally given by Gilkey and Nikčević in [105] was quite different.

Theorem 4.5   *Any 4-dimensional (para)-Kähler–Weyl curvature model can be geometrically realized by a (para)-Kähler–Weyl manifold.*

4.2   (PARA)-KÄHLER–WEYL STRUCTURES IF $m \geq 6$

We follow the discussion of Gilkey and Nikčević in [102, 103] to prove Theorem 4.1. By Theorem 1.25, we have the following decompositions into irreducible $\mathcal{U}_\pm^\star$ modules where, if $m = 4$, we omit the modules $\{W_{\pm, 5}, W_{\pm, 6}, W_{\pm, 10}\}$ and where, if $m = 6$, we omit the module $W_{\pm, 6}$ from these decompositions:

$$\mathfrak{R} = W_{\pm, 1} \oplus \cdots \oplus W_{\pm, 10}, \qquad \mathfrak{K}_{\pm, \mathfrak{R}} = W_{\pm, 1} \oplus W_{\pm, 2} \oplus W_{\pm, 3},$$
$$\mathfrak{W} = W_{\pm, 1} \oplus \cdots \oplus W_{\pm, 13}.$$

Set:

$$\mathfrak{K}_{\pm, \mathfrak{W}}^1 := \left\{ \oplus_{4 \leq i \leq 13} W_{\pm, i} \right\} \cap \mathfrak{K}_{\pm, \mathfrak{W}}.$$

Since $W_{\pm, 1} \oplus W_{\pm, 2} \oplus W_{\pm, 3}$ is a submodule of $\mathfrak{K}_{\pm, \mathfrak{W}}$, we have:

$$\mathfrak{K}_{\pm, \mathfrak{W}} = W_{\pm, 1} \oplus W_{\pm, 2} \oplus W_{\pm, 3} \oplus \mathfrak{K}_{\pm, \mathfrak{W}}^1.$$

We prove Theorem 4.1 (1) by showing $\mathfrak{K}_{\pm, \mathfrak{W}}^1 = \{0\}$ if $m \geq 6$. Suppose that $4 \leq i \leq 13$ and $i \neq 9, 13$. Since $W_{\pm, i}$ appears with multiplicity 1 in $\oplus_{4 \leq i \leq 13} W_{\pm, i}$, Lemma 1.16 shows either $W_{\pm, i} \subset \mathfrak{K}_{\pm, \mathfrak{W}}^1$ or $W_{\pm, i} \perp \mathfrak{K}_{\pm, \mathfrak{W}}^1$. By Theorem 1.25, $W_{\pm, i} \cap \mathfrak{K}_{\pm, \mathfrak{R}} = \{0\}$ for $4 \leq i \leq 10$. Consequently,

$$\mathfrak{K}_{\pm, \mathfrak{W}}^1 = \left\{ W_{\pm, 9} \oplus W_{\pm, 11} \oplus W_{\pm, 12} \oplus W_{\pm, 13} \right\} \cap \mathfrak{K}_{\pm, \mathfrak{W}}.$$

We recall the notation of Theorem 1.24. Decompose $\mathfrak{W}$ as an $\mathcal{O}$ module in the form:

$$\mathfrak{W} = \mathsf{I} \oplus S_0^2 \oplus \mathfrak{C} \oplus \mathfrak{P}.$$

Here, ρ_a provides an isomorphism from $\mathfrak{P}$ to Λ^2 and I denotes the *trivial module*. We recall the inverse isomorphism is provided by the map $\Xi : \Lambda^2 \to \mathfrak{W}$ described in Theorem 1.24:

$$\Xi(\psi)(x, y, z, w) := 2\psi(x, y)\langle z, w \rangle + \psi(x, z)\langle y, w \rangle - \psi(y, z)\langle x, w \rangle$$
$$- \psi(x, w)\langle y, z \rangle + \psi(y, w)\langle x, z \rangle. \tag{4.2.a}$$

We also recall the isomorphism Ψ from $\Lambda_\pm^2$ to $W_{\pm,9}$ given in Theorem 1.25:

$$\Psi(\psi)(x, y, z, w) := 2\langle x, J_\pm y\rangle \psi(z, J_\pm w) + 2\langle z, J_\pm w\rangle \psi(x, J_\pm y)$$
$$+\langle x, J_\pm z\rangle \psi(y, J_\pm w) + \langle y, J_\pm w\rangle \psi(x, J_\pm z) \qquad (4.2.b)$$
$$-\langle x, J_\pm w\rangle \psi(y, J_\pm z) - \langle y, J_\pm z\rangle \psi(x, J_\pm w).$$

These maps will play a crucial role in our study. Here, is an outline to the rest of this section. We will first examine $W_{\pm,11} = \Xi(\Omega_\pm \cdot \mathbb{R})$ and show the elements of $W_{\pm,11}$ do not satisfy the (para)-Kähler identity of Equation (4.1.a). Our analysis here does not depend upon the assumption that $m \geq 6$ and remains valid if $m = 4$. Then we will treat the module $W_{\pm,12} = \Xi\Lambda_{\mp,0}^2$. Here, the assumption that $m \geq 6$ is essential—we shall see presently that the 4-dimensional setting is very different. Note that $W_{\pm,9} \oplus W_{\pm,13}$ is isomorphic to two copies of $\Lambda_\pm^2$; we next examine possible submodules of this module. Again, the assumption $m \geq 6$ is essential. Finally, we will pass to the geometric setting and use Assertion (1) to establish Assertion (2).

The calculation is purely algebraic. Let $m = 2\bar{m}$. In the para-complex setting, we choose an orthonormal basis $\{e_1, \ldots, e_m\}$ for V so that:

$$\langle e_{2i-1}, e_{2i-1}\rangle = 1, \quad \langle e_{2i}, e_{2i}\rangle = -1, \quad J_+ e_{2i-1} = e_{2i}, \quad J_+ e_{2i} = e_{2i-1}.$$

In the complex setting with an inner product of signature $(2\bar{p}, 2\bar{q})$, we choose the orthonormal basis $\{e_1, \ldots, e_m\}$, so:

$$\langle e_i, e_i\rangle = \left\{ \begin{array}{l} -1 \text{ if } i \leq 2\bar{p} \\ +1 \text{ if } i > 2\bar{p} \end{array} \right\}, \quad J_- e_{2i-1} = e_{2i}, \quad \text{and} \quad J_- e_{2i} = -J_- e_{2i-1}.$$

We set $h_{ij} := \langle e_i, e_j\rangle = \pm\delta_i^j$. This vanishes if $i \neq j$, while if $i = j$, it is $+1$ or -1.

THE MODULE $W_{\pm,11}$

We use Equation (4.2.a) to verify Equation (4.1.c) by computing:

$$\Xi(\Omega_\pm)(e_1, e_4, e_3, e_1) = -\Omega_\pm(e_4, J_\pm e_3)\langle e_1, e_1\rangle = -h_{11}h_{44}.$$
$$\mp\Xi(\Omega_\pm)(e_1, e_4, J_\pm e_3, J_\pm e_1) = \pm\Omega_\pm(e_1, J_\pm J_\pm e_1)\langle e_4, J_\pm e_3\rangle = h_{11}h_{44}.$$
$$\Xi(\mathbb{R} \cdot \Omega_\pm) \not\subset \mathfrak{K}_{\pm,\mathfrak{W}}^1 \text{ if } m \geq 4.$$

THE MODULE $W_{\pm,12}$

Let $\psi_{\pm,0} := e^1 \otimes e^2 - e^2 \otimes e^1 + \nu_\pm\{e^3 \otimes e^4 - e^4 \otimes e^3\}$ where $\nu_\pm$ is chosen to ensure that $\psi_{\pm,0} \perp \Omega_\pm$. We have $J_\pm^* \psi_{\pm,0} = \mp\psi_{\pm,0}$ and thus, $\psi_{\pm,0} \in \Lambda_{\mp,0}^2$. Equation (4.2.a) yields:

$$\Xi(\psi_{\pm,0})(e_5, e_1, e_2, e_5) = -\psi_{\pm,0}(e_1, e_2)\langle e_5, e_5\rangle = -h_{55},$$
$$\mp\Xi(\psi_{\pm,0})(e_5, e_1, J_\pm e_2, J_\pm e_5) = \pm\psi_{\pm,0}(e_5, J_\pm e_5)\langle e_1, J_\pm e_2\rangle = 0, \text{ and}$$
$$W_{\pm,12} \not\subset \mathfrak{K}_{\pm,\mathfrak{W}}^1 \text{ if } m \geq 6.$$

THE MODULE $W_{\pm,9} \oplus W_{\pm,13}$

Let $\psi_{\pm} := e^1 \otimes e^3 - e^3 \otimes e^1 \pm e^2 \otimes e^4 \mp e^4 \otimes e^2$. Then $J_{\pm}^* \psi_{\pm} = \pm \psi_{\pm}$ so $\psi_{\pm} \in \Lambda_{\pm}^2$. We use Equation (4.2.a) and Equation (4.2.b) to see that:

1. $\Xi(\psi_{\pm})(e_5, e_1, e_3, e_5) = -\psi_{\pm}(e_1, e_3)\langle e_5, e_5\rangle = -h_{55}$.

2. $\Xi(\psi_{\pm})(e_5, e_1, J_{\pm}e_3, J_{\pm}e_5) = \Xi(\psi_{\pm})(e_5, e_1, e_4, e_6) = 0$.

3. $\Psi(\psi_{\pm})(e_5, e_1, e_3, e_5) = 0$.

4. $\Psi(\psi_{\pm})(e_5, e_1, J_{\pm}e_3, J_{\pm}e_5) = -\psi_{\pm}(e_1, J_{\pm}e_4)\langle e_5, J_{\pm}e_6\rangle = -h_{55}$.

5. $\Xi(\psi_{\pm})(e_5, e_6, e_1, e_4) = 0$.

6. $\Xi(\psi_{\pm})(e_5, e_6, J_{\pm}e_1, J_{\pm}e_4) = 0$.

7. $\Psi(\psi_{\pm})(e_5, e_6, e_1, e_4) = 2\langle e_5, J_{\pm}e_6\rangle\psi_{\pm}(e_1, J_{\pm}e_4) = 2h_{55}$.

8. $\Psi(\psi_{\pm})(e_5, e_6, J_{\pm}e_1, J_{\pm}e_4) = 2\langle e_5, J_{\pm}e_6\rangle\psi_{\pm}(J_{\pm}e_1, J_{\pm}J_{\pm}e_4) = \pm 2h_{55}$.

For $(a, b) \neq (0, 0)$, let $\xi(a, b) := \mathrm{Range}\{a\,\Xi + b\Psi\} \subset W_{\pm,9} \oplus W_{\pm,13}$. If $\xi(a, b) \cap \mathfrak{K}_{\pm,\mathfrak{W}}^1 \neq \{0\}$, then $\xi(a, b) \subset \mathfrak{K}_{\pm,\mathfrak{W}}^1$ by Lemma 1.16. Assertions (1)-(4) then yield $a = \mp b$, while Assertions (5)-(8) yield $b = 0$. Thus, no submodule of the form $\xi(a, b)$ intersects $\mathfrak{K}_{\pm,\mathfrak{W}}^1$. We apply Lemma 1.22 to see that every non-trivial proper submodule of $W_{\pm,9} \oplus W_{\pm,13}$ is isomorphic to $\xi(a, b)$ for some $(a, b) \neq 0$. Thus,

$$\left\{W_{\pm,9} \oplus W_{\pm,13}\right\} \cap \mathfrak{K}_{\pm,\mathfrak{W}}^1 = \{0\}.$$

Consequently, $\mathfrak{K}_{\pm,\mathfrak{W}}^1 = \{0\}$. This completes the proof of Theorem 4.1 (1).

THE GEOMETRIC SETTING

Suppose that the first de Rham cohomology group $H_{\mathrm{DeR}}^1(M)$ vanishes. Then, by Theorem 1.11, a Weyl structure (M, g, ∇) is trivial if and only if the alternating Ricci tensor ρ_a of ∇ vanishes or, equivalently, if the curvature R of ∇ is Riemannian, i.e., $R \in \mathfrak{R}$. Assertion (2) of Theorem 4.1 now follows from Assertion (1), since any element of $\mathfrak{K}_{\pm,\mathfrak{W}}$ belongs to $\mathfrak{K}_{\pm,\mathfrak{R}}$ and thus, has symmetric Ricci tensor. $\qquad\square$

4.3   (PARA)-KÄHLER–WEYL STRUCTURES IF $m = 4$

There are a number of steps in the proof of Theorem 4.2. We first show that if a (para)-Kähler–Weyl structure exists in dimension four, then it is necessarily unique. We then establish the result in the para-Hermitian context. Finally, we use analytic continuation to discuss the complex setting. We assume $m = 4$ henceforth in this chapter.

THE UNIQUENESS OF THE (PARA)-KÄHLER–WEYL STRUCTURE

We begin with a purely algebraic result.

Lemma 4.6   *Let $(V, \langle \cdot, \cdot \rangle, J_\pm)$ be a 4-dimensional (para)-Hermitian vector space. If ϕ belongs to V^*, let $\Theta_X(Y) := \phi(X)Y + \phi(Y)X - \langle X, Y \rangle \phi^\#$, where $\phi^\# \in V$ is dual to ϕ. Assume that $[\Theta_X, J_\pm] = 0$ for all X. Then $\phi = 0$.*

Proof. We first assume that $(V, \langle \cdot, \cdot \rangle, J_+)$ is a para-Hermitian vector space. Choose a hyperbolic basis $\{e_1, e_2, e_3, e_4\}$ which diagonalizes J_+:

$$J_+e_1 = e_1, \quad J_+e_2 = e_2, \quad J_+e_3 = -e_3, \quad J_+e_4 = -e_4, \quad \langle e_1, e_3 \rangle = \langle e_2, e_4 \rangle = 1.$$

Let $\phi = a_1 e^1 + a_2 e^2 + a_3 e^3 + a_4 e^4$ where $\{e^1, e^2, e^3, e^4\}$ is the corresponding dual basis for V^*. Then

$$\begin{aligned}
\Theta_{e_1}e_4 &= a_1 e_4 + a_4 e_1, & J_+\Theta_{e_1}e_4 &= -a_1 e_4 + a_4 e_1, & \Theta_{e_1}J_+e_4 &= -a_1 e_4 - a_4 e_1, \\
\Theta_{e_2}e_3 &= a_2 e_3 + a_3 e_2, & J_+\Theta_{e_2}e_3 &= -a_2 e_3 + a_3 e_2, & \Theta_{e_2}J_+e_3 &= -a_2 e_3 - a_3 e_2, \\
\Theta_{e_4}e_1 &= a_4 e_1 + a_1 e_4, & J_+\Theta_{e_4}e_1 &= a_4 e_1 - a_1 e_4, & \Theta_{e_4}J_+e_1 &= a_4 e_1 + a_1 e_4, \\
\Theta_{e_3}e_2 &= a_3 e_2 + a_2 e_3, & J_+\Theta_{e_3}e_2 &= a_3 e_2 - a_2 e_3, & \Theta_{e_3}J_+e_2 &= a_3 e_2 + a_2 e_3.
\end{aligned}$$

Equating $\Theta_{e_i}J_+e_j$ with $J_+\Theta_{e_i}e_j$ then implies $a_1 = a_2 = a_3 = a_4 = 0$ so $\phi = 0$. This establishes the Lemma in the para-Hermitian case.

Next, assume we are in the pseudo-Hermitian setting. Complexify and extend the inner product to be complex bilinear. Choose a local frame $\{Z_1, Z_2, \bar{Z}_1, \bar{Z}_2\}$ for $V \otimes_\mathbb{R} \mathbb{C}$ so

$$J_-Z_1 = \sqrt{-1}Z_1, \quad J_-Z_2 = \sqrt{-1}Z_2, \quad J_-\bar{Z}_1 = -\sqrt{-1}\bar{Z}_1, \quad J_-\bar{Z}_2 = -\sqrt{-1}\bar{Z}_2,$$
$$\langle Z_1, \bar{Z}_1 \rangle = 1, \quad \langle Z_2, \bar{Z}_2 \rangle = \varepsilon_2.$$

Take $\varepsilon_2 = +1$ if $\langle \cdot, \cdot \rangle$ has signature $(0, 4)$ and take $\varepsilon_2 = -1$ if $\langle \cdot, \cdot \rangle$ has signature $(2, 2)$. Set

$$J_+ := -\sqrt{-1}J_-, \quad e_1 := Z_1, \quad e_2 := Z_2, \quad e_3 := \bar{Z}_1, \quad e_4 := \varepsilon_2 \bar{Z}_2$$

and apply the argument given above to derive the Lemma in the pseudo-Hermitian setting where the coefficients a_i are now complex. $\square$

In the geometric setting, let $^g\nabla$ be the Levi-Civita connection of (M, g) and let $^\phi\nabla$ be a Weyl connection. Equation (1.2.q) yields the identity:

$$^\phi\nabla_X Y = {}^g\nabla_X Y + \phi(X)Y + \phi(Y)X - g(X, Y)\phi^\#.$$

Let $\phi = \phi_1 - \phi_2$ and let $\Theta_X(Y) := \phi(X)Y + \phi(Y)X - g(X, Y)\phi^\#$. Then, if $^{\phi_1}\nabla$ and $^{\phi_2}\nabla$ are two Weyl connections for (M, g), we have:

$$^{\phi_1}\nabla_X - {}^{\phi_2}\nabla_X = \Theta_X.$$

Thus, if $^{\phi_1}\nabla J_\pm = 0$ and $^{\phi_2}\nabla J_\pm = 0$, we have $[\Theta_X, J_\pm] = 0$ for all tangent vector fields X. The following result is now an immediate consequence of Lemma 4.6:

Corollary 4.7   *If $^{\phi_1}\nabla$ and $^{\phi_2}\nabla$ are two (para)-Kähler–Weyl connections on a 4-dimensional (para)-Hermitian manifold, then $\phi_1 = \phi_2$.*

A PARA-HERMITIAN EXAMPLE

Let $\{e_1, e_2, e_3, e_4\}$ be a hyperbolic basis for an inner product space $(V, \langle \cdot, \cdot \rangle)$ of signature $(2, 2)$; $\langle e_1, e_3 \rangle = \langle e_2, e_4 \rangle = 1$. We take the orientation to be given by $e^1 \wedge e^3 \wedge e^2 \wedge e^4$. The Hodge $\star$ operator is characterized by the identity:

$$\omega_1 \wedge \star \omega_2 = \langle \omega_1, \omega_2 \rangle e^1 \wedge e^3 \wedge e^2 \wedge e^4 .$$

Lemma 4.8

$$\star(e^1 \wedge e^3) = -e^2 \wedge e^4, \quad \star(e^2 \wedge e^4) = -e^1 \wedge e^3, \quad \star(e^1 \wedge e^2 \wedge e^3) = -e^2,$$
$$\star(e^1 \wedge e^2 \wedge e^4) = e^1, \quad \star(e^1 \wedge e^3 \wedge e^4) = -e^4, \quad \star(e^2 \wedge e^3 \wedge e^4) = e^3.$$

Proof. The calculation is straightforward, but it is important to be careful concerning the signs. To simplify the notation, we set $e^{i_1 \cdots i_k} := e^{i_1} \wedge \cdots \wedge e^{i_k}$. For each $\omega_2 = e^I$, there is a unique $\omega_1 = e^J$ so that $\langle e^J, e^I \rangle = \pm 1$; $\star e^I$ is then composed of the complementary indices to J and the sign is controlled by the identity $e^J \wedge \star e^I = \langle e^J, e^I \rangle e^{1324}$. We compute:

$\omega_1,$	$\omega_2,$	$\star \omega_2,$	$\omega_1 \wedge \star \omega_2 = \varepsilon e^{1324},$	$\langle \omega_1, \omega_2 \rangle = $	$\varepsilon,$
$e^{13},$	$e^{13},$	$-e^{24},$	$e^{13} \wedge (-e^{24}) = -e^{1324},$	$\langle e^{13}, e^{13} \rangle = $	$-1,$
$e^{24},$	$e^{24},$	$-e^{13},$	$e^{24} \wedge (-e^{13}) = -e^{1324},$	$\langle e^{24}, e^{24} \rangle = $	$-1,$
$e^{134},$	$e^{123},$	$-e^2,$	$e^{134} \wedge (-e^2) = +e^{1324},$	$\langle e^{134}, e^{123} \rangle = $	$+1,$
$e^{234},$	$e^{124},$	$+e^1,$	$e^{234} \wedge (+e^1) = +e^{1324},$	$\langle e^{124}, e^{234} \rangle = $	$+1,$
$e^{123},$	$e^{134},$	$-e^4,$	$e^{123} \wedge (-e^4) = +e^{1324},$	$\langle e^{123}, e^{134} \rangle = $	$+1,$
$e^{124},$	$e^{234},$	$+e^3,$	$e^{124} \wedge (+e^3) = +e^{1324},$	$\langle e^{124}, e^{234} \rangle = $	$+1.$

The Lemma now follows. □

The following example will be useful in what follows; to simplify certain expressions, set $\partial_i := \partial_{x_i}$ and $f_i := \partial_i f$.

Lemma 4.9 *Let (x^1, x^2, x^3, x^4) be the usual coordinates on $\mathbb{R}^4$. Let $f \in C^\infty(\mathbb{R}^4)$. Let the metric be determined by $g(\partial_1, \partial_3) = 1$ and by $g(\partial_2, \partial_4) = e^{2f}$. Let J_+ be the standard para-complex structure $J_+\partial_1 = \partial_1$, $J_+\partial_2 = \partial_2$, $J_+\partial_3 = -\partial_3$, and $J_+\partial_4 = -\partial_4$. Let ∇ be the Weyl connection determined by taking $\phi = \frac{1}{2} J_+ \delta_g \Omega_+$. Then $\nabla J_+ = 0$.*

Proof. The (possibly) non-zero Christoffel symbols of $^g\nabla$ are given by:

$$g(^g\nabla_{\partial_1}\partial_2, \partial_4) = g(^g\nabla_{\partial_2}\partial_1, \partial_4) = g(^g\nabla_{\partial_1}\partial_4, \partial_2) = g(^g\nabla_{\partial_4}\partial_1, \partial_2) = f_1 e^{2f},$$
$$g(^g\nabla_{\partial_3}\partial_2, \partial_4) = g(^g\nabla_{\partial_2}\partial_3, \partial_4) = g(^g\nabla_{\partial_3}\partial_4, \partial_2) = g(^g\nabla_{\partial_4}\partial_3, \partial_2) = f_3 e^{2f},$$
$$g(^g\nabla_{\partial_4}\partial_4, \partial_2) = 2 f_4 e^{2f},$$
$$g(^g\nabla_{\partial_2}\partial_2, \partial_4) = 2 f_2 e^{2f},$$

$$g({}^g\nabla_{\partial_2}\partial_4, \partial_1) = g({}^g\nabla_{\partial_4}\partial_2, \partial_1) = -f_1 e^{2f},$$

$$g({}^g\nabla_{\partial_2}\partial_4, \partial_3) = g({}^g\nabla_{\partial_4}\partial_2, \partial_3) = -f_3 e^{2f}.$$

Consequently, the (possibly) non-zero covariant derivatives are given by:

$$\begin{aligned}
&{}^g\nabla_{\partial_1}\partial_2 = {}^g\nabla_{\partial_2}\partial_1 = f_1\partial_2, \qquad &&{}^g\nabla_{\partial_1}\partial_4 = {}^g\nabla_{\partial_4}\partial_1 = f_1\partial_4,\\
&{}^g\nabla_{\partial_3}\partial_2 = {}^g\nabla_{\partial_2}\partial_3 = f_3\partial_2, \qquad &&{}^g\nabla_{\partial_3}\partial_4 = {}^g\nabla_{\partial_4}\partial_3 = f_3\partial_4,\\
&{}^g\nabla_{\partial_4}\partial_4 = 2f_4\partial_4, \qquad &&{}^g\nabla_{\partial_2}\partial_2 = 2f_2\partial_2,\\
&{}^g\nabla_{\partial_2}\partial_4 = {}^g\nabla_{\partial_4}\partial_2 = -f_1 e^{2f}\partial_3 - f_3 e^{2f}\partial_1.
\end{aligned}$$

Since ${}^g\nabla_{\partial_1}$ and ${}^g\nabla_{\partial_3}$ are diagonal, they commute with J_+ so ${}^g\nabla_{\partial_1}(J_+) = {}^g\nabla_{\partial_3}(J_+) = 0$. Thus:

$$\begin{aligned}
({}^g\nabla_{\partial_2}J_+)\partial_1 &= (\mathrm{Id} - J_+){}^g\nabla_{\partial_2}\partial_1 = (\mathrm{Id} - J_+)f_1\partial_2 = 0,\\
({}^g\nabla_{\partial_2}J_+)\partial_2 &= (\mathrm{Id} - J_+){}^g\nabla_{\partial_2}\partial_2 = (\mathrm{Id} - J_+)2f_2\partial_2 = 0,\\
({}^g\nabla_{\partial_2}J_+)\partial_3 &= (-\mathrm{Id} - J_+){}^g\nabla_{\partial_2}\partial_3 = (-\mathrm{Id} - J_+)f_3\partial_2 = -2f_3\partial_2,\\
({}^g\nabla_{\partial_2}J_+)\partial_4 &= (-\mathrm{Id} - J_+){}^g\nabla_{\partial_2}\partial_4 = (-\mathrm{Id} - J_+)e^{2f}(-f_1\partial_3 - f_3\partial_1) = 2f_3 e^{2f}\partial_1,\\
({}^g\nabla_{\partial_4}J_+)\partial_1 &= (\mathrm{Id} - J_+){}^g\nabla_{\partial_4}\partial_1 = (\mathrm{Id} - J_+)f_1\partial_4 = 2f_1\partial_4,\\
({}^g\nabla_{\partial_4}J_+)\partial_2 &= (\mathrm{Id} - J_+){}^g\nabla_{\partial_4}\partial_2 = (\mathrm{Id} - J_+)e^{2f}(-f_1\partial_3 - f_3\partial_1) = -2e^{2f}f_1\partial_3,\\
({}^g\nabla_{\partial_4}J_+)\partial_3 &= (-\mathrm{Id} - J_+){}^g\nabla_{\partial_4}\partial_3 = (-\mathrm{Id} - J_+)f_3\partial_4 = 0,\\
({}^g\nabla_{\partial_4}J_+)\partial_4 &= (-\mathrm{Id} - J_+){}^g\nabla_{\partial_4}\partial_4 = (-\mathrm{Id} - J_+)2f_4\partial_4 = 0.
\end{aligned}$$

Set $e^1 = dx^1$, $e^2 = e^f dx^2$, $e^3 = dx^3$, and $e^4 = e^f dx^4$. This is a hyperbolic basis for the cotangent space. We apply Lemma 4.8 to compute:

$$\begin{aligned}
\star\Omega_+ &= \star(-e^1 \wedge e^3 - e^2 \wedge e^4) = dx^1 \wedge dx^3 + e^{2f}dx^2 \wedge dx^4,\\
d\star\Omega_+ &= 2f_1 e^{2f}dx^1 \wedge dx^2 \wedge dx^4 - 2f_3 e^{2f}dx^2 \wedge dx^3 \wedge dx^4,\\
\delta_g\Omega_+ &= -\star d\star\Omega_+ = -2f_1 dx^1 + 2f_3 dx^3,\\
\phi &= \tfrac{1}{2}J_+\delta_g\Omega_+ = -f_1 dx^1 - f_3 dx^3, \text{ and } \phi^\# = -f_1\partial_3 - f_3\partial_1.
\end{aligned} \qquad (4.3.a)$$

Let $\Theta_{ij} := \phi(\partial_i)\partial_j + \phi(\partial_j)\partial_i - g(\partial_i, \partial_j)\phi^\# = ({}^\phi\nabla - {}^g\nabla)_{\partial_i}\partial_j$. Then:

$$\begin{aligned}
&\Theta_{11} = -2f_1\partial_1, \quad \Theta_{12} = -f_1\partial_2,\\
&\Theta_{13} = (-f_1\partial_3 - f_3\partial_1) + (f_1\partial_3 + f_3\partial_1) = 0,\\
&\Theta_{14} = -f_1\partial_4, \quad \Theta_{22} = 0, \quad \Theta_{23} = -f_3\partial_2,\\
&\Theta_{24} = e^{2f}(f_1\partial_3 + f_3\partial_1), \quad \Theta_{33} = -2f_3\partial_3, \quad \Theta_{34} = -f_3\partial_4, \quad \Theta_{44} = 0.
\end{aligned}$$

Since $\Theta(\partial_1)$ and $\Theta(\partial_3)$ are diagonal, $[\Theta(\partial_1), J_+] = [\Theta(\partial_3), J_+] = 0$. We compute:

$$
\begin{aligned}
[\Theta(\partial_2), J_+]\partial_1 &= (\mathrm{Id} - J_+)\Theta_{12} = 0, \\
[\Theta(\partial_2), J_+]\partial_2 &= (\mathrm{Id} - J_+)\Theta_{22} = 0, \\
[\Theta(\partial_2), J_+]\partial_3 &= (-\mathrm{Id} - J_+)\Theta_{23} = 2f_3\partial_2, \\
[\Theta(\partial_2), J_+]\partial_4 &= (-\mathrm{Id} - J_+)\Theta_{24} = -2f_3 e^{2f}\partial_1, \\
[\Theta(\partial_4), J_+]\partial_1 &= (\mathrm{Id} - J_+)\Theta_{14} = -2f_1\partial_4, \\
[\Theta(\partial_4), J_+]\partial_2 &= (\mathrm{Id} - J_+)\Theta_{24} = 2f_1 e^{2f}\partial_3, \\
[\Theta(\partial_4), J_+]\partial_3 &= (-\mathrm{Id} - J_+)\Theta_{34} = 0, \\
[\Theta(\partial_4), J_+]\partial_4 &= (-\mathrm{Id} - J+)\Theta_{44} = 0.
\end{aligned}
$$

We now observe that $[{}^g\nabla, J_+] + [\Theta, J_+] = 0$. Consequently, ${}^\phi\nabla J_+ = 0$.   □

PARA-KÄHLER–WEYL STRUCTURES FOR PARA-HERMITIAN METRICS

Adopt the notation of Equation (1.5.d) and let S_-^2 be the vector space of symmetric 2-cotensors ω so that $J_+^*\omega = -\omega$. Let (M, g, J_+) be the germ of a 4-dimensional para-Hermitian manifold and let P be a point of M. We can choose local coordinates that are centered at P so $J_+\partial_1 = \partial_1$, $J_+\partial_2 = \partial_2$, $J_+\partial_3 = -\partial_3$, $J_+\partial_4 = -\partial_4$, and so

$$
g = dx^1 \circ dx^3 + dx^2 \circ dx^4 + \varepsilon,
$$

where $\varepsilon \in C^\infty(S_-^2)$ satisfies $\varepsilon(0) = 0$. Theorem 4.2 in the para-Hermitian setting will follow from the following result:

Lemma 4.10   *Adopt the notation established above to define the germ of a para-Hermitian manifold $\mathcal{M}_\varepsilon$ at the origin. If $\phi := \frac{1}{2}J_+\delta\Omega_+$, then $({}^\phi\nabla J_+)(0) = 0$.*

Proof. Since only the 1-jets of ε are relevant in examining $({}^\phi\nabla J_+)(0)$, this is a linear problem and we may take $\varepsilon \in S_-^2 \otimes V^*$ and express thereby

$$
g = g_0 + x^i \varepsilon(e_i).
$$

Then $\varepsilon \to ({}^\phi\nabla J_+)(0)$ defines a linear map $\mathcal{E} : S_-^2 \otimes V^* \to \mathrm{End}\{V\} \otimes V^*$ or, equivalently,

$$
\mathcal{E} : S_-^2 \to \mathrm{Hom}\{V^*, \mathrm{End}\{V\} \otimes V^*\}.
$$

Lemma 4.9 shows that $\mathcal{E}(dx^2 \circ dx^4) = 0$. Permuting the indices $1 \leftrightarrow 2$ and $3 \leftrightarrow 4$ then yields $\mathcal{E}(dx^1 \circ dx^3) = 0$. The question is invariant under the action of the para-unitary group; we must preserve J_+ and we must preserve the inner product at the origin. Define a unitary transformation T by setting:

$$
T(e^1) = e^1 + ae^2, \quad T(e^2) = e^2, \quad T(e^3) = e^3, \quad T(e^4) = e^4 - ae^3.
$$

Then $T(e^1 \wedge e^3) = e^1 \circ e^3 + ae^2 \circ e^3$. Consequently, $\mathcal{E}(e^2 \circ e^3) = 0$. Permuting the indices $1 \leftrightarrow 2$ and $3 \leftrightarrow 4$ then yields $\mathcal{E}(e^1 \circ e^4) = 0$ as well. Since

$$S_-^2 = \mathrm{Span}\{e^1 \circ e^3, e^1 \circ e^4, e^2 \circ e^3, e^2 \circ e^4\},$$

$\mathcal{E} = 0$. This completes the proof and, as a consequence, Theorem 4.2 in the para-Hermitian setting follows. □

ANALYTIC CONTINUATION

We will derive Theorem 4.2 in the complex setting from Lemma 4.10 using analytic continuation. Let $V = \mathbb{R}^4$ with the usual basis $\{e_1, e_2, e_3, e_4\}$ and coordinates $\{x^1, x^2, x^3, x^4\}$. Let $\vec{x} := x^1 e_1 + x^2 e_2 + x^3 e_3 + x^4 e_4$. We consider:

$$\mathcal{S} := \left\{ S^2 \otimes_{\mathbb{R}} \mathbb{C} \right\} \oplus \left\{ (V^* \otimes_{\mathbb{R}} S^2) \otimes_{\mathbb{R}} \mathbb{C} \right\}.$$

Let $J_+ \in M_2(\mathbb{C})$ be a complex 2×2 matrix with $J_+^2 = \mathrm{Id}$ and $\mathrm{Tr}\{J_+\} = 0$. Let:

$$\mathcal{S}(J_+) := \{(g_0, g_1) \in \mathcal{S} : \det\{g_0 - J_+^* g_0\} \neq 0\}. \qquad (4.3.b)$$

For $(g_0, g_1) \in \mathcal{S}(J_+)$, define:

$$g(\vec{x})(X, Y) := \tfrac{1}{2}\{g_0(X, Y) - g_0(J_+ X, J_+ Y)\}$$
$$+ \sum_{i=1}^{4} x^i \cdot \tfrac{1}{2}\{g_1(e_i, X, Y) - g_1(e_i, J_+ X, J_+ Y)\}.$$

By Equation (4.3.b), this is non-degenerate at 0 and defines a complex metric on some neighborhood of 0 so $J_+^* g = -g$. Let $^g\nabla$ be the complex Levi-Civita connection:

$$^g\nabla_{\partial_i}\partial_j = \tfrac{1}{2} g^{k\ell}\{\partial_i g_{j\ell} + \partial_j g_{i\ell} - \partial_{x_\ell} g_{ij}\}\partial_k.$$

Then $^g\nabla$ is a torsion-free connection on $T_{\mathbb{C}}M := TM \otimes_{\mathbb{R}} \mathbb{C}$. The para-Kähler form is defined by setting $\Omega_+(x, y) = g(x, J_+ y)$ and we have

$$\delta\Omega_+ = \star d\Omega_+ \text{ and } \phi := \tfrac{1}{2} J_+ \delta_g \Omega_+.$$

We then use ϕ to define a *complex Weyl connection* $^\phi\nabla$ on $T_{\mathbb{C}}M$ and define a holomorphic map from $\mathcal{S}(J_+)$ to $\mathfrak{V} := M_4(\mathbb{C}) \otimes_{\mathbb{R}} V^*$ by setting:

$$\mathcal{E}(g_0, g_1; J_+) := {}^\phi\nabla(J_+)|_{\vec{x}=0}.$$

Lemma 4.11   *Let $J_+ \in M_4(\mathbb{C})$ with $J_+^2 = \mathrm{Id}$ and $\mathrm{Tr}\{J_+\} = 0$. Suppose $(g_0, g_1) \in \mathcal{S}(J_+)$.*

1. If J_+ is real and if (g_0, g_1) is real, then $\mathcal{E}(g_0, g_1; J_+) = 0$.

2. If J_+ is real and if (g_0, g_1) is complex, then $\mathcal{E}(g_0, g_1; J_+) = 0$.

3. If J_+ is complex and if (g_0, g_1) is complex, then $\mathcal{E}(g_0, g_1; J_+) = 0$.

Proof. Assertion (1) of the Lemma is simply a restatement of Lemma 4.10. We argue as follows to prove Assertion (2) of the Lemma. $\mathcal{S}(J_+)$ is an open dense subset of $\mathcal{S}$ and inherits a natural holomorphic structure thereby. Assume that J_+ is real. The map $\mathcal{E}$ is a holomorphic map from $\mathcal{S}(J_+)$ to $\mathfrak{V}$. By Assertion (1), $\mathcal{E}(g_0, g_1; J_+)$ vanishes if (g_0, g_1) is real. Thus, by the *identity theorem*, $\mathcal{E}(g_0, g_1; J_+)$ vanishes for all $(g_0, g_1) \in \mathcal{S}(J_+)$. Since we have removed the assumption that (g_0, g_1) is real, Assertion (2) of Lemma 4.11 follows.

We complete the proof of Lemma 4.11 by removing the assumption that J_+ is real. The *complex general linear group* $\mathrm{GL}_4(\mathbb{C})$ acts on the structures involved by change of basis (i.e., *conjugation*). Let $(g_0, g_1) \in \mathcal{S}(J_+)$ where J_+ is real and $\mathrm{Tr}\{J_+\} = 0$. We consider the real and complex *orbits*:

$$\mathcal{O}_{\mathbb{R}}(g_0, g_1; J_+) := \mathrm{GL}_4(\mathbb{R}) \cdot (g_0, g_1; J_+) \text{ and } \mathcal{O}_{\mathbb{C}}(g_0, g_1; J_+) := \mathrm{GL}_4(\mathbb{C}) \cdot (g_0, g_1; J_+).$$

Let $\mathcal{F}(A) := \mathcal{E}(A \cdot (g_0, g_1; J_+))$ define a holomorphic map from $\mathrm{GL}_4(\mathbb{C})$ to $\mathfrak{V}$. By Assertion (2), $\mathcal{F}$ vanishes on $\mathrm{GL}_4(\mathbb{R})$. Thus, by the identity theorem, $\mathcal{F}$ vanishes on $\mathrm{GL}_4(\mathbb{C})$ or, equivalently, $\mathcal{E}$ vanishes on the orbit space $\mathcal{O}_{\mathbb{C}}(g_0, g_1; J_+)$. Given any $J_+ \in M_4(\mathbb{C})$ with $J_+^2 = \mathrm{Id}$ and $\mathrm{Tr}\{J_+\} = 0$, we can choose $A \in \mathrm{GL}_4(\mathbb{C})$ so that $A \cdot J_+$ is real. The general case now follows from Assertion (2). $\square$

Let (M, g, J_-) be a 4-dimensional pseudo-Hermitian manifold of dimension four. Fix a point P of M. Since J_- is integrable, we may choose local coordinates (x^1, x^2, x^3, x^4) so the matrix of J_- relative to the coordinate frame $\{\partial_i\}$ is constant. Define a Weyl connection with associated 1-form given by $\phi = -\frac{1}{2} J_- \delta \Omega_-$. Only the 0-jet and the 1-jet of the metric play a role in the computation of $({}^\phi \nabla J_-)(P)$. So we may assume $g = g(g_0, g_1)$. We set $J_+ = \sqrt{-1} J_-$. We have that

$$J_+^2 = \sqrt{-1} J_- \sqrt{-1} J_- = -J_-^2 = \mathrm{Id}, \quad \mathrm{Tr}\{J_+\} = \sqrt{-1} \, \mathrm{Tr}\{J_-\} = 0,$$
$$J_+^*(g)(X, Y) = g(\sqrt{-1} J_- X, \sqrt{-1} J_- Y) = -g(J_- X, J_- Y) = -g(X, Y),$$

so $J_+^*(g) = -g$ and $(g_0, g_1) \in \mathcal{S}(J_+)$. Finally, since $J_- = -\sqrt{-1} J_+$, we have

$$\Omega_- = -\sqrt{-1} \Omega_+,$$
$$\phi_{J_-} = -\tfrac{1}{2} J_- \delta_g \Omega_- = -\tfrac{1}{2}(-\sqrt{-1} J_+) \delta_g (-\sqrt{-1} \Omega_+) = \tfrac{1}{2} J_+ \delta_g \Omega_+ = \phi_{J_+}.$$

We apply Lemma 4.11 to complete the proof of Theorem 4.2. $\square$

4.4   (PARA)-KÄHLER–WEYL LIE GROUPS IF $m = 4$

In Section 4.4, we shall establish Theorem 4.3. We treat both the para-Hermitian setting and the Hermitian setting; we do not deal with pseudo-Hermitian structures of signature $(2, 2)$. We shall describe two different Lie algebras. For generic values of the parameters, the Lie algebra in the para-complex setting is modeled on $A_{2,2} \oplus A_{2,2}$ and the Lie algebra in the complex setting is modeled on $A_{4,12}$ in the classification given by Patera *et al.* in [176]. There is a duality between these two Lie algebras which we describe as follows.

Definition 4.12

1. Let $(V, \langle \cdot, \cdot \rangle, J_+)$ be a para-Hermitian vector space of signature $(2, 2)$. Choose a hyperbolic basis $\{\Psi_1, \Psi_2, \Psi_3, \Psi_4\}$ for V and corresponding dual basis $\{\Psi^1, \Psi^2, \Psi^3, \Psi^4\}$ for V^* so that J_+ is diagonalized:

$$\langle \Psi_1, \Psi_3 \rangle = 1, \quad \langle \Psi_2, \Psi_4 \rangle = 1, \quad \langle \Psi^1, \Psi^3 \rangle = 1, \quad \langle \Psi^2, \Psi^4 \rangle = 1,$$
$$J_+ \Psi_1 = \Psi_1, \quad J_+ \Psi_2 = \Psi_2, \quad J_+ \Psi_3 = -\Psi_3, \quad J_+ \Psi_4 = -\Psi_4,$$
$$J_+ \Psi^1 = \Psi^1, \quad J_+ \Psi^2 = \Psi^2, \quad J_+ \Psi^3 = -\Psi^3, \quad J_+ \Psi^4 = -\Psi^4.$$

Let $\varepsilon_1, \tilde{\varepsilon}_1, \alpha_2, \tilde{\alpha}_2, \alpha_3, \tilde{\alpha}_3$ be real parameters. Define a bracket $[\cdot, \cdot]$ on V by setting:

$$[\Psi_1, \Psi_2] = \varepsilon_1 \Psi_1, \qquad [\Psi_1, \Psi_3] = 0, \qquad\qquad [\Psi_1, \Psi_4] = \alpha_3 \Psi_1,$$
$$[\Psi_2, \Psi_3] = -\tilde{\alpha}_3 \Psi_3, \qquad [\Psi_2, \Psi_4] = \alpha_2 \Psi_1 - \tilde{\alpha}_2 \Psi_3, \qquad [\Psi_3, \Psi_4] = \tilde{\varepsilon}_1 \Psi_3.$$

2. Let $(V, \langle \cdot, \cdot \rangle, J_-)$ be a positive definite (i.e., signature $(0, 4)$) Hermitian vector space. Choose an orthogonal basis $\{e_i\}$ for V so that:

$$\langle e_1, e_1 \rangle = 2, \quad \langle e_2, e_2 \rangle = 2, \quad \langle e_3, e_3 \rangle = 2, \quad \langle e_4, e_4 \rangle = 2,$$
$$\langle e^1, e^1 \rangle = \tfrac{1}{2}, \quad \langle e^2, e^2 \rangle = \tfrac{1}{2}, \quad \langle e^3, e^3 \rangle = \tfrac{1}{2}, \quad \langle e^4, e^4 \rangle = \tfrac{1}{2},$$
$$J_- e_1 = e_2, \quad J_- e_2 = -e_1, \quad J_- e_3 = e_4, \quad J_- e_4 = -e_3,$$
$$J_- e^1 = -e^2, \quad J_- e^2 = e^1, \quad J_- e^3 = -e^4, \quad J_- e^4 = e^3.$$

We define a complex basis $\{Z_1, Z_2, \bar{Z}_1, \bar{Z}_2\}$ for $V_{\mathbb{C}} := V \otimes_{\mathbb{R}} \mathbb{C}$ and the corresponding complex dual basis $\{Z^1, Z^2, \bar{Z}^1, \bar{Z}^2\}$ for the complex dual space $V_{\mathbb{C}}^*$ by setting:

$$Z_1 = \tfrac{1}{2}(e_1 - \sqrt{-1}e_2), Z_2 = \tfrac{1}{2}(e_3 - \sqrt{-1}e_4), Z^1 = (e^1 + \sqrt{-1}e^2), Z^2 = (e^3 + \sqrt{-1}e^4),$$
$$\bar{Z}_1 = \tfrac{1}{2}(e_1 + \sqrt{-1}e_2), \bar{Z}_2 = \tfrac{1}{2}(e_3 + \sqrt{-1}e_4), \bar{Z}^1 = (e^1 - \sqrt{-1}e^2), \bar{Z}^2 = (e^3 - \sqrt{-1}e^4).$$

Then we have

$$\langle Z_1, \bar{Z}_1 \rangle = 1, \quad \langle Z^1, \bar{Z}^1 \rangle = 1, \quad \langle Z_2, \bar{Z}_2 \rangle = 1, \quad \langle Z^2, \bar{Z}^2 \rangle = 1,$$
$$J_- Z_1 = \sqrt{-1}Z_1, \quad J_- \bar{Z}_1 = -\sqrt{-1}\bar{Z}_1, \quad J_- Z_2 = \sqrt{-1}Z_2, \quad J_- \bar{Z}_2 = -\sqrt{-1}\bar{Z}_2,$$
$$J_- Z^1 = \sqrt{-1}Z^1, \quad J_- \bar{Z}^1 = -\sqrt{-1}\bar{Z}^1, \quad J_- Z^2 = \sqrt{-1}Z^2, \quad J_- \bar{Z}^2 = -\sqrt{-1}\bar{Z}^2.$$

Let $\varepsilon_1, \alpha_2, \alpha_3$ be complex parameters. Define a complex bracket $[\cdot, \cdot]$ on $V_{\mathbb{C}}$ that arises from a corresponding real bracket on V by setting:

$$[Z_1, Z_2] = \varepsilon_1 Z_1, \qquad [Z_1, \bar{Z}_1] = 0, \qquad\qquad [Z_1, \bar{Z}_2] = \alpha_3 Z_1,$$
$$[Z_2, \bar{Z}_1] = -\bar{\alpha}_3 \bar{Z}_1, \quad [Z_2, \bar{Z}_2] = \alpha_2 Z_1 - \bar{\alpha}_2 \bar{Z}_1, \quad [\bar{Z}_1, \bar{Z}_2] = \bar{\varepsilon}_1 \bar{Z}_1.$$

3. To have a common notation, let:

para-complex setting: $\xi_1 = \Psi_1, \quad \xi_2 = \Psi_2, \quad \xi_3 = \Psi_3, \quad \xi_4 = \Psi_4, \quad \varepsilon_+ = +1,$
complex setting: $\xi_1 = Z_1, \quad \xi_2 = Z_2, \quad \xi_3 = \bar{Z}_1, \quad \xi_4 = \bar{Z}_2, \quad \varepsilon_- = \sqrt{-1}.$

The $\{\xi_i\}$ form a real (resp. complex) basis for V (resp. $V \otimes_{\mathbb{R}} \mathbb{C}$) so that

$$J_\pm \xi_1 = \varepsilon_\pm \xi_1, \quad J_\pm \xi_2 = \varepsilon_\pm \xi_2, \quad J_\pm \xi_3 = -\varepsilon_\pm \xi_3, \quad J_\pm \xi_4 = -\varepsilon_\pm \xi_4,$$
$$J_\pm \xi^1 = \varepsilon_\pm \xi^1, \quad J_\pm \xi^2 = \varepsilon_\pm \xi^2, \quad J_\pm \xi^3 = -\varepsilon_\pm \xi^3, \quad J_\pm \xi^4 = -\varepsilon_\pm \xi^4.$$

In the complex setting, we let $\tilde{\varepsilon}_1 = \bar{\varepsilon}_1$, $\tilde{\alpha}_2 = \bar{\alpha}_2$, and $\tilde{\alpha}_3 = \bar{\alpha}_3$.

Lemma 4.13

1. *The bracket $[\cdot, \cdot]$ satisfies the Jacobi identity; let G be the associated simply connected Lie group. $J_\pm$ defines an integrable left-invariant (para)-complex structure on G and $\langle \cdot, \cdot \rangle$ defines a left-invariant (para)-Hermitian metric on $(G, J_\pm)$.*

2. *Use Theorem 4.2 to define a corresponding left-invariant (para)-Kähler–Weyl connection ∇. The associated alternating Ricci tensor of this connection is given by:*

$$\rho_a = \tilde{\alpha}_2 \varepsilon_1 \xi^1 \wedge \xi^2 + \tilde{\alpha}_2 \alpha_3 \xi^1 \wedge \xi^4 - \alpha_2 \tilde{\alpha}_3 \xi^2 \wedge \xi^3 + \alpha_2 \tilde{\varepsilon}_1 \xi^3 \wedge \xi^4.$$

Proof. We verify the bracket $[\cdot, \cdot]$ satisfies the Jacobi identity:

$$[[\xi_1, \xi_2], \xi_3] + [[\xi_2, \xi_3], \xi_1] + [[\xi_3, \xi_1], \xi_2] = \varepsilon_1 [\xi_1, \xi_3] - \tilde{\alpha}_3 [\xi_3, \xi_1] + 0 = 0,$$

$$[[\xi_1, \xi_2], \xi_4] + [[\xi_2, \xi_4], \xi_1] + [[\xi_4, \xi_1], \xi_2]$$
$$= \varepsilon_1 [\xi_1, \xi_4] + [\alpha_2 \xi_1 - \tilde{\alpha}_2 \xi_3, \xi_1] - \alpha_3 [\xi_1, \xi_2] = \varepsilon_1 \alpha_3 \xi_1 + 0 - \alpha_3 \varepsilon_1 \xi_1 = 0,$$

$$[[\xi_1, \xi_3], \xi_4] + [[\xi_3, \xi_4], \xi_1] + [[\xi_4, \xi_1], \xi_3] = 0 + \tilde{\varepsilon}_1 [\xi_3, \xi_1] - \alpha_3 [\xi_1, \xi_3] = 0,$$

$$[[\xi_2, \xi_3], \xi_4] + [[\xi_3, \xi_4], \xi_2] + [[\xi_4, \xi_2], \xi_3]$$
$$= -\tilde{\alpha}_3 [\xi_3, \xi_4] + \tilde{\varepsilon}_1 [\xi_3, \xi_2] - [\alpha_2 \xi_1 - \tilde{\alpha}_2 \xi_3, \xi_3] = -\tilde{\alpha}_3 \tilde{\varepsilon}_1 \xi_3 + \tilde{\varepsilon}_1 \tilde{\alpha}_3 \xi_3 + 0 = 0.$$

Let $V_+ := \mathrm{Span}\{\xi_1, \xi_2\}$ and $V_- := \mathrm{Span}\{\xi_3, \xi_4\}$ be the ± 1 (resp. $\pm\sqrt{-1}$) eigenspaces of J_+ (resp. J_-). Then

$$[V_+, V_+] \subset V_+ \text{ and } [V_-, V_-] \subset V_-$$

so $J_\pm$ is integrable by Theorem 1.13; this establishes Assertion (1).

We work in the para-complex setting for the moment to establish Assertion (2). Express the para-Kähler form as:

$$\Omega_+ := -(\xi^1 \wedge \xi^3 + \xi^2 \wedge \xi^4).$$

By Equation (1.1.c), we have $d\xi^i(\Psi_j, \Psi_k) = -\xi^i([\Psi_j, \Psi_k])$. Consequently:

$$d\xi^1 = -\varepsilon_1 \xi^1 \wedge \xi^2 - \alpha_3 \xi^1 \wedge \xi^4 - \alpha_2 \xi^2 \wedge \xi^4, \quad d\xi^2 = 0,$$
$$d\xi^3 = \tilde{\alpha}_3 \xi^2 \wedge \xi^3 + \tilde{\alpha}_2 \xi^2 \wedge \xi^4 - \tilde{\varepsilon}_1 \xi^3 \wedge \xi^4, \quad d\xi^4 = 0.$$

We apply Lemma 4.8 to compute $\star$ relative to this basis simply replacing e^i by ξ^i. Since $\delta = -\star d \star$ and $\star \Omega_+ = -\Omega_+$, we have:

$$\delta \Omega_+ = -\star d \star \Omega_+ = -\star d(\xi^1 \wedge \xi^3 + \xi^2 \wedge \xi^4)$$
$$= \star\{(\varepsilon_1 \xi^1 \wedge \xi^2 + \alpha_3 \xi^1 \wedge \xi^4 + \alpha_2 \xi^2 \wedge \xi^4) \wedge \xi^3\}$$
$$+ \star \{\xi^1 \wedge (\tilde{\alpha}_3 \xi^2 \wedge \xi^3 + \tilde{\alpha}_2 \xi^2 \wedge \xi^4 - \tilde{\varepsilon}_1 \xi^3 \wedge \xi^4)\}$$
$$= \tilde{\alpha}_2 \xi^1 + (-\varepsilon_1 - \tilde{\alpha}_3)\xi^2 - \alpha_2 \xi^3 + (\tilde{\varepsilon}_1 + \alpha_3)\xi^4,$$

$$dJ_+\delta\Omega_+ = d\{\tilde{\alpha}_2 \xi^1 + (-\varepsilon_1 - \tilde{\alpha}_3)\xi^2 + \alpha_2 \xi^3 - (\tilde{\varepsilon}_1 + \alpha_3)\xi^4\}$$
$$= -\tilde{\alpha}_2 \varepsilon_1 \xi^1 \wedge \xi^2 - \tilde{\alpha}_2 \alpha_3 \xi^1 \wedge \xi^4 - \tilde{\alpha}_2 \alpha_2 \xi^2 \wedge \xi^4$$
$$+ \alpha_2 \tilde{\alpha}_3 \xi^2 \wedge \xi^3 + \alpha_2 \tilde{\alpha}_2 \xi^2 \wedge \xi^4 - \alpha_2 \tilde{\varepsilon}_1 \xi^3 \wedge \xi^4.$$

By Theorem 4.2, $\phi_+ = \frac{1}{2}J_+\delta\Omega_+$. By Equation (1.2.s), $\rho_a = -2d\phi_+$. Consequently, we have $\rho_a = -dJ_+\delta\Omega_+$ and the desired result follows. In the complex setting, $J_- = \sqrt{-1}J_+$ and $\Omega_- = \sqrt{-1}\Omega_+$ and ρ_a is unchanged as

$$\phi_- = -\frac{1}{2}J_-\delta\Omega_- = -\frac{1}{2}\sqrt{-1}\sqrt{-1}J_+\delta\Omega_+ = \phi_+. \qquad \square$$

Let $SO(\Lambda^2_+, 0)$ and $SO(\Lambda^2_\pm)$ denote the special orthogonal group of these inner product spaces; in the positive definite setting, $SO(\Lambda^2_{+,0}) \approx S^3$ and $SO(\Lambda^2_-) \approx S^1$.

Lemma 4.14   *Let V be a real vector space of dimension four.*

1. *Let $(\langle \cdot, \cdot \rangle, J_+)$ be a para-Hermitian structure on V. Then $\Lambda^2_{-,0}$ has signature $(2, 1)$ and Λ^2_+ has signature $(1, 1)$. Every orbit of the action of $\mathcal{U}_+$ on $\Lambda^2_{-,0} \oplus \Lambda^2_+$ contains a representative perpendicular to $\Psi^1 \wedge \Psi^3 - \Psi^2 \wedge \Psi^4$.*

2. *Let $(\langle \cdot, \cdot \rangle, J_-)$ be a positive definite Hermitian structure on V. Then the natural action of the unitary group $\mathcal{U}_-$ on $\Lambda^2_{+,0} \oplus \Lambda^2_-$ defines a surjective group homomorphism π from $\mathcal{U}_-$ to $SO(\Lambda^2_{+,0}) \oplus SO(\Lambda^2_-)$.*

Proof. The para-Kähler form is given by $\Omega_+ = -\Psi^1 \wedge \Psi^3 - \Psi^2 \wedge \Psi^4$. We define an orthogonal basis $\{\theta_1, \theta_2, \theta_3\}$ for $\Lambda^2_{-,0}$ and an orthogonal basis $\{\theta_4, \theta_5\}$ for Λ^2_+ by setting:

$$\theta_1 := \Psi^1 \wedge \Psi^3 - \Psi^2 \wedge \Psi^4, \quad \theta_2 := \Psi^1 \wedge \Psi^4 + \Psi^2 \wedge \Psi^3, \quad \theta_3 := \Psi^1 \wedge \Psi^4 - \Psi^2 \wedge \Psi^3,$$
$$\theta_4 := \Psi^1 \wedge \Psi^2 + \Psi^3 \wedge \Psi^4, \quad \theta_5 := \Psi^1 \wedge \Psi^2 - \Psi^3 \wedge \Psi^4.$$

We show that $\Lambda^2_{-,0}$ has signature $(2, 1)$ and that Λ^2_+ has signature $(1, 1)$ by computing:

$$\langle \theta_1, \theta_1 \rangle = -2, \quad \langle \theta_2, \theta_2 \rangle = -2, \quad \langle \theta_3, \theta_3 \rangle = 2, \quad \langle \theta_4, \theta_4 \rangle = 2, \quad \langle \theta_5, \theta_5 \rangle = -2,$$
$$\langle \theta_i, \theta_j \rangle = 0 \text{ for } i \neq j \,.$$

Define an element of $\mathcal{U}_+$ by setting:

$$T_\theta \Psi^1 := \quad \cos\theta \, \Psi^1 + \sin\theta \, \Psi^2, \quad T_\theta \Psi^3 := \quad \cos\theta \, \Psi^3 + \sin\theta \, \Psi^4,$$
$$T_\theta \Psi^2 := -\sin\theta \, \Psi^1 + \cos\theta \, \Psi^2, \quad T_\theta \Psi^4 := -\sin\theta \, \Psi^3 + \cos\theta \, \Psi^4.$$

We then have that:

$$T_\theta\theta_1 = \cos(2\theta)\theta_1 + \sin(2\theta)\theta_2, \quad T_\theta\theta_2 = -\sin(2\theta)\theta_1 + \cos(2\theta)\theta_2, \quad T_\theta\theta_i = \theta_i \ \text{ for } i \geq 3 \,.$$

We complete the proof of Assertion (1) by performing an appropriate rotation in the plane spanned by $\{\theta_1, \theta_2\}$ to eliminate the coefficient of θ_1.

We now establish Assertion (2). We have

$$\Lambda^2_{+,0} \otimes_\mathbb{R} \mathbb{C} = \mathrm{Span}_\mathbb{C}\{Z^1 \wedge \bar{Z}^1 - Z^2 \wedge \bar{Z}^2, Z^1 \wedge \bar{Z}^2, \bar{Z}^1 \wedge Z^2\},$$
$$\Lambda^2_- \otimes_\mathbb{R} \mathbb{C} = \mathrm{Span}_\mathbb{C}\{Z^1 \wedge Z^2, \bar{Z}^1 \wedge \bar{Z}^2\} \,.$$

We construct generators for the unitary group. First define $T_\theta \in \mathcal{U}_-$ by:

$$T_\theta(e^1) = \cos\theta e^1 - \sin\theta e^2, \quad T_\theta(e^2) = \sin\theta e^1 + \cos\theta e^2,$$
$$T_\theta(e^3) = \cos\theta e^3 - \sin\theta e^4, \quad T_\theta(e^4) = \sin\theta e^3 + \cos\theta e^4,$$
$$T_\theta(Z^1) = e^{\sqrt{-1}\theta} Z^1, \quad T_\theta(Z^2) = e^{\sqrt{-1}\theta} Z^2, \quad T_\theta(\bar{Z}^1) = e^{-\sqrt{-1}\theta}\bar{Z}^1,$$
$$T_\theta(\bar{Z}^2) = e^{-\sqrt{-1}\theta}\bar{Z}^2, \quad T_\theta(Z^1 \wedge \bar{Z}^1) = Z^1 \wedge \bar{Z}^1, \quad T_\theta(Z^2 \wedge \bar{Z}^2) = Z^2 \wedge \bar{Z}^2,$$
$$T_\theta(Z^1 \wedge \bar{Z}^2) = Z^1 \wedge \bar{Z}^2, \quad T_\theta(\bar{Z}^1 \wedge Z^2) = \bar{Z}^1 \wedge Z^2,$$
$$T_\theta(Z^1 \wedge Z^2) = e^{2\sqrt{-1}\theta} Z^1 \wedge Z^2, T_\theta(\bar{Z}^1 \wedge \bar{Z}^2) = e^{-2\sqrt{-1}\theta}\bar{Z}^1 \wedge \bar{Z}^2.$$

Thus, T_θ is the identity on $\Lambda^2_{+,0} \otimes_\mathbb{R} \mathbb{C}$ and acts as a rotation through an angle of 2θ on the underlying real vector space Λ^2_-. This realizes the $\mathcal{SO}(2)$ on Λ^2_- action and reduces the proof of Assertion (2) to examining the action on $\Lambda^2_{+,0}$. Next, define $\tilde{T}_\theta \in \mathcal{U}_-$ by setting:

$$\tilde{T}_\theta(e^1) = \cos\theta e^1 - \sin\theta e^2, \quad \tilde{T}_\theta(e^2) = \quad \sin\theta e^1 + \cos\theta e^2,$$
$$\tilde{T}_\theta(e^3) = \cos\theta e^3 + \sin\theta e^4, \quad \tilde{T}_\theta(e^4) = -\sin\theta e^3 + \cos\theta e^4,$$

$$\tilde{T}_\theta(Z^1) = e^{\sqrt{-1}\theta} Z^1, \quad \tilde{T}_\theta(Z^2) = e^{-\sqrt{-1}\theta} Z^2, \quad \tilde{T}_\theta(\bar{Z}^1) = e^{-\sqrt{-1}\theta} \bar{Z}^1,$$
$$\tilde{T}_\theta(\bar{Z}^2) = e^{\sqrt{-1}\theta} \bar{Z}^2, \quad \tilde{T}_\theta(Z^1 \wedge \bar{Z}^1) = Z^1 \wedge \bar{Z}^1, \quad \tilde{T}_\theta(Z^2 \wedge \bar{Z}^2) = Z^2 \wedge \bar{Z}^2,$$
$$\tilde{T}_\theta(Z^1 \wedge \bar{Z}^2) = e^{2\sqrt{-1}\theta} Z^1 \wedge \bar{Z}^2, \quad \tilde{T}_\theta(\bar{Z}^1 \wedge Z^2) = e^{-2\sqrt{-1}\theta} \bar{Z}^1 \wedge Z^2,$$
$$\tilde{T}_\theta(Z^1 \wedge Z^2) = Z^1 \wedge Z^2, \quad \tilde{T}_\theta(\bar{Z}^1 \wedge \bar{Z}^2) = \bar{Z}^1 \wedge \bar{Z}^2.$$

Consequently, $\tilde{T}_\theta$ is the identity on $\Lambda^2_- \otimes_{\mathbb{R}} \mathbb{C}$, is the identity on $\{Z^1 \wedge \bar{Z}^1 - Z^2 \wedge \bar{Z}^2\} \cdot \mathbb{C}$, and acts as a rotation through an angle of 2θ on the real vector space

$$\operatorname{Span}_{\mathbb{R}}\{Z^1 \wedge \bar{Z}^2 + Z^1 \wedge \bar{Z}^2, \sqrt{-1}(Z^1 \wedge \bar{Z}^2 - Z^1 \wedge \bar{Z}^2)\}.$$

We complete the proof of Assertion (2) by defining $\check{T}_\phi \in \mathcal{U}_-$:

$$\check{T}_\phi(e^1) = \cos\phi e^1 + \sin\phi e^3, \quad \check{T}_\phi(e^2) = \cos\phi e^2 + \sin\phi e^4,$$
$$\check{T}_\phi(e^3) = \cos\phi e^3 - \sin\phi e^1, \quad \check{T}_\phi(e^4) = \cos\phi e^4 - \sin\phi e^2,$$
$$\check{T}_\phi(Z^1) = \cos\phi Z^1 + \sin\phi Z^2, \quad \check{T}_\phi(Z^2) = \cos\phi Z^2 - \sin\phi Z^1,$$
$$\check{T}_\phi(\bar{Z}^1) = \cos\phi \bar{Z}^1 + \sin\phi \bar{Z}^2, \quad \check{T}_\phi(\bar{Z}^2) = \cos\phi \bar{Z}^2 - \sin\phi \bar{Z}^1,$$
$$\check{T}_\phi(\sqrt{-1}(Z^1 \wedge \bar{Z}^1 - Z^2 \wedge \bar{Z}^2)) = (\cos^2\phi - \sin^2\phi)\sqrt{-1}(Z^1 \wedge \bar{Z}^1 - Z^2 \wedge \bar{Z}^2)$$
$$+ 2\cos\phi\sin\phi\sqrt{-1}(Z^1 \wedge \bar{Z}^2 - \bar{Z}^1 \wedge Z^2),$$
$$\check{T}_\phi(\sqrt{-1}(Z^1 \wedge \bar{Z}^2 - \bar{Z}^1 \wedge Z^2)) = -2\cos\phi\sin\phi\sqrt{-1}(Z^1 \wedge \bar{Z}^1 - Z^2 \wedge \bar{Z}^2)$$
$$+ (\cos^2\phi - \sin^2\phi)\sqrt{-1}(Z^1 \wedge \bar{Z}^2 - \wedge\bar{Z}^1 \wedge Z^2),$$
$$\check{T}_\phi(Z^1 \wedge \bar{Z}^2 + \bar{Z}^1 \wedge Z^2) = Z^1 \wedge Z^2 + \bar{Z}^1 \wedge Z^2,$$
$$\check{T}_\phi(Z^1 \wedge Z^2) = Z^1 \wedge Z^2, \quad \text{and} \quad \check{T}_\phi(\bar{Z}^1 \wedge \bar{Z}^2) = \bar{Z}^1 \wedge \bar{Z}^2.$$

Consequently, $\check{T}_\phi$ is the identity on Λ^2_-, is the identity on $(Z^1 \wedge \bar{Z}^2 + \bar{Z}^1 \wedge Z^2) \cdot \mathbb{C}$, and acts as a rotation through an angle of 2ϕ on

$$\operatorname{Span}_{\mathbb{R}}\left\{\sqrt{-1}(Z^1 \wedge \bar{Z}^1 - Z^2 \wedge \bar{Z}^2), \sqrt{-1}(Z^1 \wedge \bar{Z}^2 - \bar{Z}^1 \wedge Z^2)\right\}.$$

The elements $\{\tilde{T}_\theta, \check{T}_\phi\}$ generate $\mathcal{SO}(\Lambda^2_{+,0})$ and fix Λ^2_-. Assertion (2) follows. $\qquad\square$

THE PROOF OF THEOREM 4.3

The set of geometrically representable tensors is invariant under the action of the structure group $\mathcal{U}^\star_+$. Let $\Xi \in \Lambda^2_{-,0} \oplus \Lambda^2_+$. By Lemma 4.14 (1), we may assume:

$$\Xi = \mu_{12}\Psi^1 \wedge \Psi^2 + \mu_{14}\Psi^1 \wedge \Psi^4 + \mu_{23}\Psi^2 \wedge \Psi^3 + \mu_{34}\Psi^3 \wedge \Psi^4;$$

in other words, there is no $\Psi^1 \wedge \Psi^3 - \Psi^2 \wedge \Psi^4$ term. We consider the Lie group of Lemma 4.13 and set $\alpha_2 = \tilde{\alpha}_2 = 1$. The remaining parameters are then determined; we complete the proof by taking:

$$\varepsilon_1 = \mu_{12}, \alpha_3 = \mu_{14}, \tilde{\alpha}_3 = -\mu_{23}, \tilde{\varepsilon}_1 = \mu_{34}.$$

Next we treat the complex setting. Let $\Xi = \Xi_{+,0} + \Xi_- \in \Lambda^2_{+,0} \oplus \Lambda^2_-$. We wish to show Ξ is geometrically representable. The question of representability is invariant under the action of $\mathcal{U}_-$ or, equivalently, by Lemma 4.14 (2), under the action of $SO(\Lambda^2_{+,0}) \oplus SO(\Lambda^2_-)$. Thus, only the norms $|\Xi_{+,0}|$ and $|\Xi_-|$ are relevant in establishing Theorem 4.3 (2). We apply Lemma 4.13 (2) to compute ρ_a and to see

$$|\Xi_{+,0}|^2 = 2|\alpha_2|^2|\alpha_3|^2 \quad \text{and} \quad |\Xi_-|^2 = 2|\alpha_2|^2|\varepsilon_1|^2.$$

If we set $\alpha_2 = 1$, we may then adjust α_3 and ε_1 to obtain arbitrary norms for $\Xi_{+,0}$ and Ξ_- and complete the proof. $\qquad\square$

4.5   (PARA)-KÄHLER–WEYL TENSORS IF $m = 4$

This section is devoted to the proof of Theorem 4.4. We restate the short exact sequence of Equation (4.1.b):

$$0 \to \mathfrak{K}_{\pm,\mathfrak{R}} \to \mathfrak{K}_{\pm,\mathfrak{W}} \to \rho_a\{\mathfrak{K}_{\pm,\mathfrak{W}}\} \to 0.$$

If η is an irreducible $\mathcal{U}^*_\pm$ module and if ξ is a submodule of $\otimes^4 V^*$, let $n_\eta(\xi)$ be the multiplicity with which η appears in the decomposition of ξ given in Lemma 1.16. Then:

$$n_\eta(\mathfrak{K}_{\pm,\mathfrak{W}}) = n_\eta(\mathfrak{K}_{\pm,\mathfrak{R}}) + n_\eta(\rho_a\{\mathfrak{K}_{\pm,\mathfrak{W}}\}).$$

We apply Theorem 1.25. If $\eta = W_{i,\pm}$ for $i \in \{1, 2, 3, 6, 7, 8, 10\}$, then $n_\eta(\Lambda^2) = 0$. Consequently, we may estimate:

$$\left\{ \begin{array}{ll} 1 & \text{if } i = 1, 2, 3 \\ 0 & \text{if } i = 6, 7, 8, 10 \end{array} \right\} = n_\eta(\mathfrak{K}_{\pm,\mathfrak{R}}) \le n_\eta(\mathfrak{K}_{\pm,\mathfrak{R}}) + n_\eta(\rho_a(\mathfrak{K}_{\pm,\mathfrak{W}}))$$

$$= n_\eta(\mathfrak{K}_{\pm,\mathfrak{W}}) \le n_\eta(\mathfrak{K}_{\pm,\mathfrak{R}}) + n_\eta(\Lambda^2) = \left\{ \begin{array}{ll} 1 & \text{if } i = 1, 2, 3 \\ 0 & \text{if } i = 6, 7, 8, 10 \end{array} \right\} + 0.$$

This shows that all of the inequalities in the above display must have been equalities and determines $n_\eta(\mathfrak{K}_{\pm,\mathfrak{W}})$ for these representations. Thus, only the multiplicities of the representations $\{\Omega_\pm \cdot \mathbb{R}, \Lambda^2_{\mp,0}, \Lambda^2_\pm\}$ are at issue. The analysis performed previously in Section 4.2 shows that the module $\Omega_\pm \cdot \mathbb{R}$ plays no role. Furthermore, the analysis performed above shows that:

$$0 \le n_{\Lambda^2_{\mp,0}}(\mathfrak{K}_{\pm,\mathfrak{W}}) \le 1 \text{ and } 0 \le n_{\Lambda^2_\pm}(\mathfrak{K}_{\pm,\mathfrak{W}}) \le 1.$$

We must consider the module $\eta := \Lambda^2_\pm$ or $\eta := \Lambda^2_{\mp,0}$ in dimension $m = 4$. We have that $n_\eta(\mathfrak{K}_{\pm,\mathfrak{W}}) \le 1$. Thus, if we can exhibit a non-trivial element of

$$W_{\pm,12} \cap \mathfrak{K}_{\pm,\mathfrak{W}} \text{ or } \{W_{\pm,9} \oplus W_{\pm,13}\} \cap \mathfrak{K}_{\pm,\mathfrak{W}},$$

we will have $n_\eta(\mathfrak{K}_{\pm,\mathfrak{W}}) = 1$. The existence of such a non-trivial element follows from Theorem 4.3 in the Hermitian and the para-Hermitian settings. To deal with the pseudo-Hermitian setting in signature $(2, 2)$, we complexify. Let $(V, \langle \cdot, \cdot \rangle, J_-)$ be a Hermitian vector space of signature $(0, 4)$ and let $A \in \eta(V, \langle \cdot, \cdot \rangle, J_-) \subset \mathfrak{K}_{-,\mathfrak{W}}(V, \langle \cdot, \cdot \rangle, J_-)$ for $\eta \approx \Lambda^2_{+,0}$ or $\eta \approx \Lambda^2_-$. Let $V_{\mathbb{C}} := V \otimes_{\mathbb{R}} \mathbb{C}$. Extend $\langle \cdot, \cdot \rangle$, J_-, and A to be complex bilinear, complex linear, and complex multilinear, respectively. Let:

$$V_{2,2} := \mathrm{Span}_{\mathbb{R}}\{\sqrt{-1}e_1, \sqrt{-1}e_2, e_3, e_4\}.$$

Let $\Re(z)$ and $\Im(z)$ denote the real and imaginary parts, respectively, of a complex number. Then $(\langle \cdot, \cdot \rangle, J_-)$ restricts to a pseudo-Hermitian almost complex structure on $V_{2,2}$ of signature $(2, 2)$. Note that

$$\Re(A|_{V_{2,2}}) \in \eta(V_{2,2}, \langle \cdot, \cdot \rangle, J_-) \cap \mathfrak{K}_{-,\mathfrak{W}}(V_{2,2}),$$
$$\Im(A|_{V_{2,2}}) \in \eta(V_{2,2}, \langle \cdot, \cdot \rangle, J_-) \cap \mathfrak{K}_{-,\mathfrak{W}}(V_{2,2}).$$

Since $A|_{V_{2,2}} \neq 0$, at least one of these tensors is non-trivial and the desired conclusion follows. This completes the proof of Theorem 4.4. $\qquad\square$

4.6 REALIZABILITY OF (PARA)-KÄHLER–WEYL TENSORS IF $m = 4$

This section is devoted to the proof of Theorem 4.5. We shall first show that any element of $\mathfrak{K}_{\pm,\mathfrak{W}}$ is geometrically realizable by the germ of a suitable structure.

Let $(V, \langle \cdot, \cdot \rangle, J_\pm)$ be a (para)-Hermitian vector space. Extend $J_\pm$ to an integrable (para)-complex structure on TV by identifying the tangent bundle TV with the trivial bundle $V \times V$ over V. Let $\{e_i\}$ be a basis for V and let $\{x^i\}$ be the dual system of coordinates on V; we normalize the coordinate system so that

$$J_\pm \partial_{x_1} = \partial_{x_2}, \qquad J_\pm \partial_{x_2} = \pm\partial_{x_1}, \qquad J_\pm \partial_{x_3} = \partial_{x_4}, \qquad J_\pm \partial_{x_4} = \pm\partial_{x_3},$$
$$J_\pm dx^1 = \pm dx^2, \qquad J_\pm dx^2 = dx^1, \qquad J_\pm dx^3 = \pm dx^4, \qquad J_\pm dx^4 = dx^3.$$

Let $\varepsilon_{ij} := \langle e_i, e_j \rangle$; we shall also let ε denote $\langle \cdot, \cdot \rangle$. Let $\theta \in S^2_\pm \otimes S^2$. Set:

$$g_\theta = (\varepsilon_{ij} + \theta_{ijk\ell}x^k x^\ell)dx^i \circ dx^j.$$

Since $g_\theta(0) = \varepsilon$, g_θ is non-degenerate near 0 and defines a pseudo-Riemannian metric on some neighborhood $\mathcal{O}$ of 0 in V. Since $\theta \in S^2_\pm \otimes S^2$ and since $J^*_\pm\varepsilon = \mp\varepsilon$, $J^*_\pm g_\theta = \mp g_\theta$. Thus, $(g_\theta, J_\pm)$ defines the germ of a (para)-Hermitian manifold at 0 in V.

By Theorem 4.2, there is a unique Weyl connection $^\theta\nabla$ so that $(\mathcal{O}, J_\pm, g_\theta, {}^\theta\nabla)$ is a (para)-Kähler–Weyl manifold. Let $\Theta(\theta) := \nabla R(0)$. Since $g_\theta = \varepsilon + O(|\vec{x}|^2)$, the first derivatives of the metric play no role at 0 and Θ defines an equivariant linear map:

$$\Theta : S^2_\pm \otimes S^2 \to \mathfrak{K}_{\pm,\mathfrak{W}}.$$

We wish to show Θ is surjective. By Theorem 1.25 and Theorem 4.4, there are $\mathcal{U}_{\pm}^{\star}$ module isomorphisms

$$\mathfrak{K}_{\pm,\mathfrak{W}} \approx \mathfrak{K}_{\pm,\mathfrak{R}} \oplus \Lambda_{\mp,0}^2 \oplus \Lambda_{\pm}^2 \text{ and } \mathfrak{K}_{\pm,\mathfrak{R}} \approx \mathbf{1} \oplus S_{\mp,0}^2 \oplus W_{\pm,3} \qquad (4.6.a)$$

where the five modules $\{\mathbf{1}, S_{\mp,0}^2, W_{\pm,3}, \Lambda_{\mp,0}^2, \Lambda_{\pm}^2\}$ which appear in Equation (4.6.a) are irreducible and inequivalent $\mathcal{U}_{\pm}^{\star}$ modules. We will complete the proof by showing

$$\mathfrak{K}_{\pm,\mathfrak{R}} \subset \text{Range}\{\Theta\} \text{ and } \rho_a : \text{Range}\{\Theta\} \to \Lambda_{\mp,0}^2 \oplus \Lambda_{\pm}^2 \to 0.$$

THE MODULE $\mathfrak{K}_{\pm,\mathfrak{R}}$

We begin our discussion with the following observation:

Lemma 4.15   *If* $\theta \in S_{\pm}^2 \otimes S^2$, *let* $g_\theta := (\varepsilon_{ij} + \theta_{ijk\ell} x^k x^\ell) dx^i \circ dx^j$.

$$\mathcal{R}(\theta)(x, y, z, w) := \theta(x, z, y, w) + \theta(y, w, x, z) - \theta(x, w, y, z) - \theta(y, z, x, w) \text{ and }$$

$$\mathcal{K}_{\pm}(\theta)(x, y, z) := 2\{\theta(x, J_{\pm}y, z, e_\ell) + \theta(y, J_{\pm}z, x, e_\ell) + \theta(z, J_{\pm}x, y, e_\ell)\} x^\ell.$$

Then $^{g_\theta}R(0)(x, y, z, w) = \mathcal{R}(\theta)(x, y, z, w)$ *and* $d\Omega_{\pm}^{g_\theta}(x, y, z) = 2\mathcal{K}_{\pm}(\theta)(x, y, z)$. *Furthermore,* $\theta \in \ker\{\mathcal{K}_{\pm}\}$ *if and only if* g_θ *is a (para)-Kähler metric.*

Proof. We apply the Koszul formula of Equation (1.2.k) to see the Christoffel symbols of the Levi-Civita connection are given by:

$$^{g_\theta}\Gamma_{ijk} = \tfrac{1}{2}(\partial_i g_{\theta,jk} + \partial_j g_{\theta,ik} - \partial_k g_{\theta,ij}) = x^\ell(\theta_{jki\ell} + \theta_{ikj\ell} - \theta_{ijk\ell}).$$

Since $^{g_\theta}\Gamma_{ink} = O(|\vec{x}|)$ and since $g_\theta = \varepsilon + O(|\vec{x}|)$, we may raise indices to see

$$^{g_\theta}\Gamma_{ij}{}^n = g_\theta^{nk}\, {}^{g_\theta}\Gamma_{ijk} = \varepsilon^{nk}\, {}^{g_\theta}\Gamma_{ijk} + O(|\vec{x}|^2).$$

This permits us to determine $^{g_\theta}R(0)$ by computing:

$$
\begin{aligned}
^{g_\theta}R_{ijk}{}^p &= \partial_i{}^{g_\theta}\Gamma_{jk}{}^p - \partial_j{}^{g_\theta}\Gamma_{ik}{}^p + O(|\vec{x}|) = \varepsilon^{p\ell}\{\partial_i{}^{g_\theta}\Gamma_{jk\ell} - \partial_j{}^{g_\theta}\Gamma_{ik\ell}\} + O(|\vec{x}|), \\
^{g_\theta}R_{ijk\ell} &= \partial_i{}^{g_\theta}\Gamma_{jk\ell} - \partial_j{}^{g_\theta}\Gamma_{ik\ell} + O(|\vec{x}|) \\
&= \theta_{k\ell ji} + \theta_{j\ell ki} - \theta_{jk\ell i} - \theta_{k\ell ij} - \theta_{i\ell kj} + \theta_{ik\ell j} + O(|\vec{x}|) \\
&= \theta_{ikj\ell} + \theta_{j\ell ik} - \theta_{jki\ell} - \theta_{i\ell jk} + O(|\vec{x}|).
\end{aligned}
$$

Since $\theta \in S_{\pm}^2 \otimes S^2$, $\theta(x, J_{\pm}y, z, w) = -\theta(y, J_{\pm}x, z, w)$. We complete the proof by computing:

$$\Omega_{\pm}^{g_\theta} = \tfrac{1}{2}\sum_{i,j}\theta(e_i, J_{\pm}e_j, e_k, e_\ell)x^k x^\ell dx^i \wedge dx^j,$$

$$
\begin{aligned}
d\Omega_{\pm}^{g_\theta} &= \sum_{i,j,k}\theta(e_i, J_{\pm}e_j, e_k, e_\ell)x^\ell dx^i \wedge dx^j \wedge dx^k \\
&= 2\sum_{i<j<k}\{\theta(e_i, J_{\pm}e_j, e_k, e_\ell) + \theta(e_j, J_{\pm}e_k, e_i, e_\ell) + \theta(e_k, J_{\pm}e_i, e_j, e_\ell)\}x^\ell \\
&\qquad \times dx^i \wedge dx^j \wedge dx^k.
\end{aligned}
$$

$\square$

Adopt the notation of Definition 4.12. By Equation (1.2.p) and Equation (1.4.b),

$$\mathfrak{R} \subset S^2(\Lambda^2(V^*)) \text{ and } \mathfrak{K}_{\pm,\mathfrak{R}} \subset S^2(\Lambda_\pm^2).$$

Let $\xi^{ijk\ell} := (\xi^i \wedge \xi^j) \circ (\xi^k \wedge \xi^\ell)$. After taking into account the Bianchi identity, we have:

$$\mathfrak{K}_{+,\mathfrak{R}} \subset \text{Span}\{\xi^{1313}, \xi^{1414}, \xi^{2323}, \xi^{2424}, \xi^{1323}, \xi^{1424}, \xi^{1314}, \xi^{2324}, \xi^{1324} + \xi^{1423}\}.$$

There are nine elements in this basis; this is in accordance with the dimension count (see Tricerri and Vanhecke [200] in the complex setting and Brozos-Vázquez, Gilkey, and Nikčević [37] in the para-complex setting) that $\dim(\mathbb{1}) = 1$, $\dim(S_{\mp,0}^2) = 3$, and $\dim(W_{\pm,3}) = 5$.

We consider the following example. Let

$$\langle \xi_1, \xi_3 \rangle = \langle \xi_2, \xi_4 \rangle = 1 \text{ and } \theta = \tfrac{1}{2}(\xi^1 \circ \xi^3) \otimes (\xi^1 \circ \xi^3) \in S_\pm^2 \otimes S^2.$$

The metric g_θ is a real (para)-Kähler metric that takes the form of the product metric of $M_1 \times M_2$, where M_1 is a *(para)-Riemann surface* and where M_2 is flat; we can also verify directly from Lemma 4.15 that the (para)-Kähler form vanishes since $\mathcal{K}_\pm(\theta)$ is a 3-form that is supported on $\text{Span}\{e_1, e_2\}$. Furthermore, an application of Lemma 4.15 permits us to compute the curvature tensor and show

$$\mathcal{R}(\theta) = \xi^{1313} \in \text{Range}\{\mathcal{K}_\pm\}. \tag{4.6.b}$$

There is an algebra acting that will be central to our treatment. Let

$$\text{End}_\pm = \{T \in \text{End}\{V\} : T J_\pm = J_\pm T\}.$$

The vector spaces $S_\pm^2 \otimes S^2$, $\mathfrak{R}$, and $\Lambda^3(V^*) \otimes V^*$ are modules over this algebra and the maps $\theta \to \mathcal{R}(\theta)$ and $\theta \to \mathcal{K}_\pm(\theta)$ of Lemma 4.15 are $\text{End}_\pm$ module morphisms. In particular, $\ker\{\mathcal{K}_\pm\}$ is an $\text{End}_\pm$ module. On $\ker\{\mathcal{K}_\pm\}$, we have $\Theta = \mathcal{R}$ since $\phi_\theta = \pm\tfrac{1}{2}J_\pm\delta\Omega_\pm = 0$. Thus:

$$\mathcal{R} : \ker\{\mathcal{K}_\pm\} \to \mathfrak{K}_{\pm,\mathfrak{R}} = \mathbb{1} \oplus S_{\mp,0}^2 \oplus W_{\pm,3}.$$

We shall always complexify when considering J_-. Let

$$T(e^1) = e^1 + ae^2, \quad T(e^2) = e^2, \quad T(e^3) = e^3 + \tilde{a}e_4, \quad T(e^4) = e^4.$$

In the para-complex setting, we let $a, \tilde{a} \in \mathbb{R}$ while in the complex setting, we let $\tilde{a} = \bar{a} \in \mathbb{C}$. Then $T \in \text{End}_\pm$. We apply T to the element of Equation (4.6.b) to conclude

$$\begin{aligned} T(\xi^{1313}) &= \xi^{1313} + 2a\xi^{1323} + 2\tilde{a}\xi^{1314} + a^2\xi^{2323} + \tilde{a}^2\xi^{1414} + 2a\tilde{a}(\xi^{1324} + \xi^{1423}) \\ &\quad + 2a^2\tilde{a}\xi^{2324} + 2a\tilde{a}^2\xi^{1424} + 2a^2\tilde{a}^2\xi^{2424} \in \text{Range}\{\mathcal{K}_\pm\}. \end{aligned} \tag{4.6.c}$$

In the para-complex setting, we take $\tilde{a} = \pm a$. Since $\xi^{1313} \in \text{Range}\{\mathcal{K}_\pm\}$, we have

$$2a\xi^{2313} \pm 2a\xi^{1314} + O(a^2) \in \text{Range}\{\mathcal{K}_\pm\} \text{ so}$$
$$\xi^{2313} + O(a) \in \text{Range}\{\mathcal{K}_+\} \text{ and } \xi^{1314} + O(a) \in \text{Range}\{\mathcal{K}_+\}.$$

Since Range$\{\mathcal{K}_+\}$ is a linear subspace of a finite-dimensional vector space, it is closed. Thus, we may take the limit as $a \to 0$ to conclude

$$\xi^{1323} \in \text{Range}\{\mathcal{K}_+\} \text{ and } \xi^{1314} \in \text{Range}\{\mathcal{K}_+\} .$$

In the complex setting, we take $a = t$ and $\tilde{a} = t$ or $a = \sqrt{-1}t$ and $\tilde{a} = -\sqrt{-1}t$ to conclude

$$2t\{\xi^{2313} + \xi^{1314}\} + O(t^2) \in \text{Range}\{\mathcal{K}_\pm\}, \text{ and}$$
$$2t\sqrt{-1}(\xi^{2313} - \xi^{1314}) + O(t^2) \in \text{Range}\{\mathcal{K}_\pm\} \text{ so}$$
$$\xi^{2313} \in \text{Range}\{\mathcal{K}_-\} \text{ and } \xi^{1314} \in \text{Range}\{\mathcal{K}_-\} .$$

Permuting the indices $1 \leftrightarrow 2$ and $3 \leftrightarrow 4$ defines an element of $\text{End}_\pm$ and shows

$$\xi^{2424} \in \text{Range}\{\mathcal{K}_\pm\}, \quad \xi^{1424} \in \text{Range}\{\mathcal{K}_\pm\}, \quad \xi^{2324} \in \text{Range}\{\mathcal{K}_\pm\} .$$

Consequently Equation (4.6.c) yields:

$$a^2\xi^{2323} + 2a\tilde{a}(\xi^{1323} + \xi^{1423}) \in \text{Range}\{\mathcal{K}_\pm\} + \tilde{a}^2\xi^{1414} .$$

A similar argument now shows

$$\{\xi^{2323}, \xi^{1323} + \xi^{1423}, \xi^{1414}\} \subset \text{Range}\{\mathcal{K}_\pm\} \text{ so } \text{Range}\{\mathcal{K}_\pm\} = \mathfrak{K}_{\pm,\mathfrak{R}} .$$

THE MODULE $\Lambda^2_{\mp,0} \oplus \Lambda^2_\pm$

In the study of $\mathbf{1}$ and $S^2_{\mp,0}$, we used the symmetric Ricci tensor ρ_s. We use the alternating Ricci tensor to study $\Lambda^2_{\mp,0}$ and $\Lambda^2_\pm$. We begin by studying the para-complex setting. We adopt the notation of Lemma 4.9. Let

$$J_+\partial_1 = \partial_1, \; J_+\partial_2 = \partial_2, \; J_+\partial_3 = -\partial_3, \; J_+\partial_4 = -\partial_4, \quad g(\partial_1, \partial_3) = 1, \quad g(\partial_2, \partial_4) = e^{2f} .$$

We computed in Equation (4.3.a) that $\phi = -f_1dx^1 - f_3dx^3$ and consequently,

$$\rho_a = -2d\phi = 2\{-f_{12}dx^1 \wedge dx^2 - f_{14}dx^1 \wedge dx^4 + f_{23}dx^2 \wedge dx^3 - f_{34}dx^3 \wedge dx^4\},$$
$$J_+\rho_a = 2\{-f_{12}dx^1 \wedge dx^2 + f_{14}dx^2 \wedge dx^3 - f_{23}dx^1 \wedge dx^4 - f_{34}dx^3 \wedge dx^4\} .$$

Since $J_+\rho_a$ is not equal to $\pm\rho_a$, ρ_a has components in both $\Lambda^2_{-,0}$ and Λ^2_+; this provides the desired example in this instance.

To study the complex setting, we consider the example:

$$J_-\partial_1 = \partial_2, \quad J_-\partial_2 = -\partial_1, \quad J_-\partial_3 = \partial_4, \quad J_-\partial_4 = -\partial_3,$$
$$g(\partial_1, \partial_1) = g(\partial_2, \partial_2) = 1, \quad g(\partial_3, \partial_3) = g(\partial_4, \partial_4) = \varepsilon e^{2f} .$$

We then set

$$Z_1 = \tfrac{1}{2}(\partial_1 - \sqrt{-1}\partial_2), \quad Z_2 = \tfrac{1}{2}(\partial_3 - \sqrt{-1}\partial_4), \quad Z_3 = \bar{Z}_1, \quad Z_4 = \bar{Z}_2,$$
$$Z^1 = dx^1 + \sqrt{-1}dx^2, \quad Z^2 = dx^3 + \sqrt{-1}dx^4, \quad Z^3 = \bar{Z}^1, \quad Z^4 = \bar{Z}^2,$$

and mimic the proof of Lemma 4.9. Let $f_i := Z_i(f)$. We compute:

$$\langle Z^1, Z^3 \rangle = 1, \quad \langle Z^2, Z^4 \rangle = e^{2f},$$

$$\Omega_- = -\sqrt{-1}\{Z^1 \wedge Z^3 + \varepsilon e^{2f} Z^2 \wedge Z^4\},$$

$$\star \Omega_- = \sqrt{-1}\{Z^1 \wedge Z^3 + \varepsilon e^{2f} Z^2 \wedge Z^4\},$$

$$d \star \Omega_- = 2\varepsilon e^{2f} \sqrt{-1}\{f_1 Z^1 \wedge Z^2 \wedge Z^4 - f_3 Z^2 \wedge Z^3 \wedge Z^4\},$$

$$\delta_g \Omega_- = \varepsilon \sqrt{-1}\{-2 f_1 Z^1 + 2 f_3 Z^3\},$$

$$\phi_- = -\tfrac{1}{2} J_- \delta_g \Omega_- = \varepsilon \{-f_1 Z^1 - f_3 Z^3\},$$

$$\rho_a = 2\varepsilon \{-f_{12} Z^1 \wedge Z^2 - f_{14} Z^1 \wedge Z^4 + f_{23} Z^2 \wedge Z^3 - f_{34} Z^3 \wedge Z^4\}.$$

Since

$$J_- Z^1 = \sqrt{-1} Z^1, \quad J_- Z^2 = \sqrt{-1} Z^2,$$
$$J_- Z^3 = -\sqrt{-1} Z^3, \quad J_- Z^4 = -\sqrt{-1} Z^4,$$

$Z^1 \wedge Z^2 \in \Lambda^2_{-,0}$ and $Z_1 \wedge Z^4 \in \Lambda^2_+$. Consequently, $J_- \rho_a$ has components in both $\Lambda^2_{-,0}$ and Λ^2_+. This shows $\Lambda^2_{\mp,0} \oplus \Lambda^2_\pm \subset \text{Range}\{\Theta\}$. This completes the proof of Theorem 4.5. $\qquad\square$

Bibliography

[1] R. Abounasr, A. Belhaj, J. Rasmussen, and E. H. Saidi, "Superstring theory on pp waves with ADE geometries", *J. Phys. A* **39** (2006), 2797–2841. DOI: 10.1088/0305-4470/39/11/015. 31

[2] J. Adams, "Vector fields on spheres", *Ann. of Math.* **75** (1962), 603–632. DOI: 10.2307/1970213. 50

[3] Z. Afifi, "Riemann extensions of affine connected spaces", *Quart. J. Math., Oxford Ser. (2)* **5** (1954), 312–320. DOI: 10.1093/qmath/5.1.312. 29, 33, 34, 35, 36, 66

[4] A. Alcolado, A. MacDougall, A. Coley, and S. Hervik, "4D neutral signature VSI and CSI spaces", *J. Geom. Phys.* **62** (2012), 594–603. DOI: 10.1016/j.geomphys.2011.04.012. 25

[5] D. V. Alekseevsky, V. Cortés, A. S. Galaev, and T. Leistner, "Cones over pseudo-Riemannian manifolds and their holonomy", *J. Reine Angew. Math.* **635** (2009), 23–69. DOI: 10.1515/CRELLE.2009.075. 29

[6] I. Anderson and G. Thompson, "The inverse problem of the calculus of variations for ordinary differential equations", *Mem. Amer. Math. Soc.* **98** (1992), 1–110. 60

[7] G. Andreoli, "Parallelismi trasporti rigidi, riferimenti locali nelle V_2", *Ann. Scuola Norm. Sup. Pisa Cl. Sci. (2)* **1** (1932), 315–332. https://eudml.org/doc/82835 14

[8] V. Apostolov, "Generalized Golberg-Sachs theorems for pseudo-Riemannian four manifolds", *J. Geom. Phys.* **27** (1998), 185–198. DOI: 10.1016/S0393-0440(97)00075-2. 49, 79, 83

[9] T. Arias-Marco and O. Kowalski, "Classification of locally homogeneous affine connections with arbitrary torsion on 2-dimensional manifolds", *Monatsh. Math* **153** (2008), 1–18. DOI: 10.1007/s00605-007-0494-0. 62, 70

[10] M. Barros and A. Romero, "Indefinite Kähler manifolds", *Math. Ann.* **261** (1982), 55–62. DOI: 10.1007/BF01456410. 48

[11] L. Bérard Bergery and A. Ikemakhen, "Sur L'holonomie des variétés pseudo-riemanniennes de signature (n, n)", *Bull. Soc. Math. France* **125** (1997), 93–114. https://eudml.org/doc/87759 29

[12] A. L. Besse, "Einstein manifolds", *Ergebnisse der Mathematik und ihrer Grenzgebiete. 3. Folge* **10**. Springer-Verlag, Berlin, 1987. DOI: 10.1007/978-3-540-74311-8. 9, 10

[13] L. Bianchi, "Vorlesungen über differentialgeometrie", Teubner (Leipzig) (1899). 11

[14] W. Blaschke, *Vorlesungen über Differentialgeometrie und geometrische Grundlagen von Einsteins Relativitätstheorie. II. Affine Differentialgeometrie*, Springer, Berlin, 1923. DOI: 10.1007/978-3-642-47392-0. 11

[15] N. Blažić and N. Bokan, "Compact Riemann surfaces with the skew–symmetric Ricci tensor", *Izv. Vyssh. Uchebn. Zaved. Mat.* **9** (1994), 8–12. 60, 61

[16] N. Blažić, N. Bokan, and P. Gilkey, "A note on Osserman Lorentzian manifolds", *Bull. London Math. Soc.* **29** (1997), 227–230. DOI: 10.1112/S0024609396002238. 50, 55, 77

[17] N. Blažić, N. Bokan, and Z. Rakić, "Osserman pseudo-Riemannian manifolds of signature (2, 2)", *J. Aust. Math. Soc.* **71** (2001), 367–395. DOI: 10.1017/S1446788700003001. 50, 80, 83, 84

[18] N. Blažić, N. Bokan, and Z. Rakić, "Foliation of a dynamically homogeneous neutral manifold", *J. Math. Phys.* **39** (1998), 6118–6124. DOI: 10.1063/1.532617. 85

[19] N. Blažić, P. Gilkey, S. Nikčević, and U. Simon, "The spectral geometry of the Weyl conformal curvature tensor", *Banach Center Publ.* **69** (2005), 195–203. DOI: 10.4064/bc69-0-15. 51

[20] N. Blažić, P. Gilkey, S. Nikčević, and U. Simon, "Algebraic theory of affine curvature tensors", *Arch. Math.* (Brno) **42** (2006), suppl. 147–168. https://eudml.org/doc/249816 15, 17

[21] N. Blažić, P. Gilkey, S. Nikčević, and I. Stavrov, "Curvature structure of self-dual 4-manifolds", *Int. J. Geom. Methods Mod. Phys.* **5** (2008), 1191–1204. DOI: 10.1142/S0219887808003259. 80

[22] N. Bokan, "On the complete decomposition of curvature tensors of Riemannian manifolds with symmetric connection", *Rend. Circ. Mat. Palermo* **39** (1990), 331–380. DOI: 10.1007/BF02844767. 20

[23] N. Bokan, P. Gilkey, and U. Simon, "Geometry of differential operators on Weyl manifolds", *Proc. R. Soc. London. Ser. A* **453** (1997), 2527–2536. DOI: 10.1098/rspa.1997.0134. 15

[24] N. Bokan, P. Gilkey, and U. Simon, "Asymptotic spectra for Weyl geometries", *Romanian conference on geometry*, An. Stiint. Univ. Al. I. Cuza Iasi. Mat. (N.S.) **42** (1996), 59–71. 15

[25] A. Bonome, R. Castro, E. García-Río, L. Hervella, and Y. Matsushita, "Null holomorphically flat indefinite almost Hermitian manifolds", *Illinois J. Math.* **39** (1995), 635–660. http://projecteuclid.org/DPubS?service=UI&version=1.0&verb=Display&handle=euclid.ijm/1255986270 48

[26] A. Bonome, R. Castro, E. García-Río, L. Hervella, and R. Vázquez-Lorenzo, "Nonsymmetric Osserman indefinite Kähler manifolds", *Proc. Amer. Math. Soc.* **126** (1998), 2763–2769. DOI: 10.1090/S0002-9939-98-04659-0. 50, 84

[27] A. Bonome, R. Castro, E. García-Río, L. Hervella, and R. Vázquez-Lorenzo, "On the paraholomorphic sectional curvature of almost para-Hermitian manifolds", *Houston J. Math.* **24** (1998), 277–300. 48, 86

[28] H. W. Brinkmann, "Einstein spaces which are mapped conformally on each other", *Math. Ann.* **94** (1925), 119–145. DOI: 10.1007/BF01208647. 31

[29] M. Brozos-Vázquez, E. García-Río, and P. Gilkey, "Relating the curvature tensor and the complex Jacobi operator of an almost Hermitian manifold", *Adv. Geom.* **8** (2008), 353–365. DOI: 10.1515/ADVGEOM.2008.023. 47, 101, 102

[30] M. Brozos-Vázquez, E. García-Río, P. Gilkey, and L. Hervella, "Geometric realizability of covariant derivative Kähler tensors for almost pseudo-Hermitian and almost para-Hermitian manifolds", *Ann. Mat. Pura Appl.* **191** (2012), 487–502. DOI: 10.1007/s10231-011-0192-3. 22, 23

[31] M. Brozos-Vázquez, E. García-Río, P. Gilkey, S. Nikčević, and R. Vázquez-Lorenzo, "The geometry of Walker manifolds", *Synthesis Lectures on Mathematics and Statistics* **5**, Morgan & Claypool Publ., Williston, VT, 2009. DOI: 10.2200/S00197ED1V01Y200906MAS005. 31

[32] M. Brozos-Vázquez, E. García-Río, P. Gilkey, and R. Vázquez-Lorenzo, "Homogeneous 4-dimensional Kähler Weyl Structures", accepted for publication in *Results in Mathematics*. 19, 105, 107

[33] M. Brozos-Vázquez, E. García-Río, and R. Vázquez-Lorenzo, "Conformally Osserman 4-dimensional manifolds whose conformal Jacobi operators have complex eigenvalues", *Proc. R. Soc. Lond. Ser. A* **462** (2006), 1425–1441. DOI: 10.1098/rspa.2005.1621. 51, 78

[34] M. Brozos-Vázquez, P. Gilkey, and E. Merino, "Geometric realizations of Kähler and of para-Kähler curvature models", *Int. J. Geom. Methods Mod. Phys.* **7** (2010), 505–515. DOI: 10.1142/S0219887810004403. 19, 28

[35] M. Brozos-Vázquez, P. Gilkey, and S. Nikčević, "Geometric realizations of affine Kähler curvature models", *Results Math.* **59** (2011), 507–521. DOI: 10.1007/s00025-011-0105-1. 20

[36] M. Brozos-Vázquez, P. Gilkey, and S. Nikčević: "The structure of the space of affine Kähler curvature tensors as a complex module", *Int. J. Geom. Methods Mod. Phys.* **8** (2011), 1849–1868. DOI: 10.1142/S0219887811005981. 20

[37] M. Brozos-Vázquez, P. Gilkey, and S. Nikčević, "Geometric realizations of curvature tensors", *ICP Advanced Texts in Mathematics* **6**, Imperial College Press, London, 2012. DOI: 10.1142/9781848167421. 18, 19, 21, 24, 25, 28, 125

[38] M. Brozos-Vázquez, P. Gilkey, S. Nikčević, and R. Vázquez-Lorenzo, "Geometric realizations of para-Hermitian curvature models", *Results Math.* **56** (2009), 319–333. DOI: 10.1007/s00025-009-0403-z. 20

[39] R. L. Bryant, "Bochner-Kähler metrics", *J. Amer. Math. Soc.* **14** (2001), 623–715. DOI: 10.1090/S0894-0347-01-00366-6. 29, 37, 79, 83

[40] G. Calvaruso, "Homogeneous structures on three-dimensional Lorentzian manifolds", *J. Geom. Phys.* **57** (2007), 1279-1291. DOI: 10.1016/j.geomphys.2006.10.005. 107

[41] G. Calvaruso, "Three-dimensional Ivanov-Petrova manifolds", *J. Math. Phys.* **50** (2009), 063509, 12 pp. DOI: 10.1063/1.3152607. 52, 56

[42] E. Calviño-Louzao, E. García-Río, P. Gilkey, and R. Vázquez-Lorenzo, "Higher dimensional Osserman metrics with non-nilpotent Jacobi operators", *Geom. Dedicata* **156** (2012), 151–163. DOI: 10.1007/s10711-011-9595-y. 50

[43] E. Calviño-Louzao, E. García–Río, P. Gilkey, and R. Vázquez-Lorenzo, "The geometry of modified Riemannian extensions", *Proc. R. Soc. Lond. Ser. A.* **465** (2009), 2023–2040. DOI: 10.1098/rspa.2009.0046. 29, 32, 38

[44] E. Calviño-Louzao, E. García-Río, and R. Vázquez-Lorenzo, "Four-dimensional Osserman Ivanov Petrova metrics of neutral signature", *Classical Quantum Gravity* **24** (2007), 2343–2355. DOI: 10.1088/0264-9381/24/9/012. 50, 52, 73, 76

[45] E. Calviño-Louzao, E. García-Río, and R. Vázquez-Lorenzo, "Riemann extensions of torsion-free connections with degenerate Ricci tensor", *Canad. J. Math.* **62** (2010), 1037–1057. DOI: 10.4153/CJM-2010-059-2. 29, 69, 70, 76

[46] E. Canfes, "On generalized recurrent Weyl spaces and Wong's conjecture", *Differ. Geom. Dyn. Syst.* **8** (2006), 34–42. 15

[47] É. Cartan, "Sur les variétés á connexion affine, et la théorie de la relativité généralisée (première partie)", *Ann. Sci. École Norm. Sup.* **40** (1923), 325–412. 11

[48] É. Cartan, "Sur les varietes a connexion projective", *Bull. Soc. Math. France* **52** (1924), 205–241. 8

[49] É. Cartan, "Sur une classe remarquable d'espaces de Riemann, I", *Bull. Soc. Math. France* **54** (1926), 214–216. 15, 41

[50] É. Cartan, "Sur une classe remarquable d'espaces de Riemann, II", *Bull. Soc. Math. France* **55** (1927), 114–134. 15, 41

[51] J. Cendán-Verdes, E. García-Río and M. E. Vázquez-Abal, "On the semi-Riemannian structure of the tangent bundle of a two-point homogeneous space", *Riv. Mat. Univ. Parma* **3** (1994), 253–270. 34

[52] M. Chaichi, E. García-Río, and Y. Matsushita, "Curvature properties of four-dimensional Walker metrics", *Classical Quantum Gravity* **22** (2005), 559–577. DOI: 10.1088/0264-9381/22/3/008. 29

[53] Q. S. Chi, "A curvature characterization of certain locally rank-one symmetric spaces", *J. Differential Geom.* **28** (1988), 187–202. http://projecteuclid.org/DPubS?service= UI&version=1.0&verb=Display&handle=euclid.jdg/1214442277 50, 55, 77, 79

[54] E. Christoffel, "Über die Transformation der homogenen Differentialausdrücke 2ten Grades.", *Borchardt J.* **70** (1869), 46–70. 8, 14

[55] A. Chudecki and M. Przanowski, "From hyperheavenly spaces to Walker and Osserman spaces I", *Classical Quantum Gravity* **25** (2008), no. 14, 145010, 18 pp. DOI: 10.1088/0264-9381/25/14/145010. 50, 85

[56] A. Chudecki and M. Przanowski, "From hyperheavenly spaces to Walker and Osserman spaces II", *Classical Quantum Gravity* **25** (2008), no. 23, 235019, 22 pp. DOI: 10.1088/0264-9381/25/23/235019. 50, 85

[57] A. Clebsch, "Über die simultane integration linearer partieller differentialgleichungen", *J. Reine. Angew. Math.* **65** (1866), 257–268. DOI: 10.1515/crll.1866.65.257. 18

[58] A. Coley, R. Milson, V. Pravda, and A. Pravdová, "Vanishing scalar invariant spacetimes in higher dimensions", *Classical Quantum Gravity* **21** (2004), 5519–5542. DOI: 10.1088/0264-9381/21/23/014. 25

[59] A. Cortés-Ayaso, J. C.Díaz-Ramos, and E. García-Río, "Four-dimensional manifolds with degenerate self-dual Weyl curvature operator", *Ann. Global Anal. Geom.* **34** (2008), 185–193. DOI: 10.1007/s10455-007-9101-9. 15, 49, 85

[60] V. Cruceanu, P. Fortuny, and P. M. Gadea, "A survey on paracomplex geometry", *Rocky Mountain J. Math.* **26** (1996), 83–115. DOI: 10.1216/rmjm/1181072105. 17, 18, 48

[61] G. Darboux, "Sur le problème de Pfaff", *C. R.* **94** (1882), 835–837; *Darb. Bull. (2)* **6** (1982), 14–36; 49–68. https://eudml.org/doc/85135 3

[62] J. Davidov and O. Muskarov, "Self-dual Walker metrics with two-step nilpotent Ricci operator", *J. Geom. Phys.* **57** (2006), 157–165. DOI: 10.1016/j.geomphys.2006.02.007. 84

[63] F. Deahna, "Über die Bedingungen der Integrabilität linearer Differentialgleichungen erster Ordnung zwischen einer beliebigen Anzahl veränderlicher Grössen", *J. Reine Angew. Math.* **20** (1840), 340–349. DOI: 10.1515/crll.1840.20.340. 18

[64] A. Derdzinski, "Self-dual Kähler manifolds and Einstein manifolds of dimension four", *Compos. Math.* **49** (1983), 405–433. 49, 79

[65] A. Derdzinski, "Curvature-homogeneous indefinite Einstein metrics in dimension four: the diagonalizable case", *Recent advances in Riemannian and Lorentzian geometries (Baltimore, MD, 2003)*, 21–38, *Contemp. Math.*, **337**, Amer. Math. Soc., Providence, RI, 2003. DOI: 10.1090/conm/337/06049. 83

[66] A. Derdzinski, "Connections with skew-symmetric Ricci tensor on surfaces", *Results Math.* **52** (2008), 223–245. DOI: 10.1007/s00025-008-0307-3. 39, 60, 61

[67] A. Derdzinski, "Non-Walker self-dual neutral Einstein four-manifolds of Petrov type III", *J. Geom. Anal.* **19** (2009), 301–357. DOI: 10.1007/s12220-008-9066-3. 85

[68] A. Derdzinski and W. Roter, "Walker's theorem without coordinates", *J. Math. Phys.* **47** (2006), 062504, 8 pp. DOI: 10.1063/1.2209167. 29, 31

[69] J. C. Díaz-Ramos and E. García-Río, "A note on the structure of algebraic curvature tensors", *Linear Algebra Appl.* **382** (2004), 271–277. DOI: 10.1016/j.laa.2003.12.044. 20

[70] J. C. Díaz-Ramos, E. García-Río, and R. Vázquez-Lorenzo, "New examples of Osserman metrics with nondiagonalizable Jacobi operators", *Differential Geom. Appl.* **24** (2006), 433–442. DOI: 10.1016/j.difgeo.2006.02.006. 50, 85

[71] J. C. Díaz-Ramos, E. García-Río, and R. Vázquez-Lorenzo, "Four-dimensional Osserman metrics with nondiagonalizable Jacobi operators", *J. Geom. Anal.* **16** (2006), 39–52. DOI: 10.1007/BF02930986. 38, 50, 84

[72] A. Di Scala and L. Vezzoni, "Gray identities, canonical connection, and integrability", *Proc. Edinb. Math. Soc.* **53** (2010), 657–674. DOI: 10.1017/S0013091509000157. 102

[73] A. Di Scala, J. Lauret, and L. Vezzoni, "Quasi-Kähler Chern-flat manifolds and complex 2-step nilpotent Lie algebras", *Ann. Sc. Norm. Super. Pisa Cl. Sci.* **11** (2012), 41–60. 102

[74] M. P. do Carmo, "Geometria riemanniana", *Projeto Euclides* **10**. Instituto de Matematica Pura e Aplicada, Rio de Janeiro, 1979. 41

[75] V. Dryuma, "The Riemann extensions in theory of differential equations and their applications", *Mat. Fiz. Anal. Geom.* **10** (2003), 307–325. http://www.mathnet.ru/php/archive.phtml?wshow=paper&jrnid=jmag&paperid=253&option_lang=eng 29

[76] M. Dunajski, "Paraconformal geometry of nth-order ODEs, and exotic holonomy in dimension four", *J. Geom. Phys.* **56** (2006), 1790–1809. DOI: 10.1016/j.geomphys.2005.10.007. 29

[77] L. Eisenhart, "Non-Riemannian geometry", American Mathematical Society Colloquium Publications **8**, *Amer. Math. Soc.*, Providence, RI, 1964. 10, 14

[78] L. Eisenhart, *Riemannian Geometry*, Princeton University Press, Princeton, N.J., 1949. http://press.princeton.edu/titles/486.html 10

[79] B. Fiedler, "Determination of the structure of algebraic curvature tensors by means of Young symmetrizers", *Sém. Lothar. Combin.* **48** (2002), Art. B48d, 20 pp. 20

[80] B. Fine, P. Kirk, and E. Klassen, "A local analytic splitting of the holonomy map on flat connections", *Math. Ann.* **299** (1994), 171–189. DOI: 10.1007/BF01459778. 29

[81] G. Frobenius, "Über das Pfaffsche probleme", *Borchardt J.* **82** (1876), 230–315. http://www.degruyter.com/view/j/crll.1877.issue-82/crll.1877.82.230/crll.1877.82.230.xml 18

[82] A. Frölicher and A. Nijenhuis, "Theory of vector valued differential forms. Part I.", *Indag. Math.* **18** (1956), 338–360. DOI: 10.1007/978-3-540-77054-1_13. 18

[83] A. Frölicher and A. Nijenhuis, "Invariance of vector form operations under mappings", *Comment. Math. Helv.* **34** (1960), 227–248. DOI: 10.1007/BF02565938. 18

[84] T. Fukami, "Invariant tensors under the real representation of unitary groups and their application", *J. Math. Soc. Japan* **10** (1958), 135–144. DOI: 10.2969/jmsj/01020135. 24

[85] P. M. Gadea and A. Montesinos-Amilibia, "Spaces of constant para-holomorphic sectional curvature", *Pacific J. Math.* **136** (1989), 85–101. DOI: 10.2140/pjm.1989.136.85. 48

[86] E. García-Río, P. Gilkey, M. E. Vázquez-Abal, and R. Vázquez-Lorenzo, "Four-dimensional Osserman metrics of neutral signature", *Pacific J. Math.* **244** (2010), 21–36. DOI: 10.2140/pjm.2010.244.21. 80, 82, 86

[87] E. García-Río, A. Haji-Badali, and R. Vázquez-Lorenzo, "Lorentzian three-manifolds with special curvature operators", *Classical Quantum Gravity* **25** (2008), 015003, 13pp. DOI: 10.1088/0264-9381/25/1/015003. 56

[88] E. García-Río, D. N. Kupeli, and M. E. Vázquez-Abal, "On a problem of Osserman in Lorentzian geometry", *Differential Geom. Appl.* **7** (1997), 85–100. DOI: 10.1016/S0926-2245(96)00037-X. 50, 55, 77

[89] E. García-Río, D. N. Kupeli, M. E. Vázquez-Abal, and R. Vázquez-Lorenzo, "Affine Osserman connections and their Riemann extensions", *Differential Geom. Appl.* **11** (1999), 145–153. DOI: 10.1016/S0926-2245(99)00029-7. 29, 49, 51

[90] E. García-Río, D. N. Kupeli, and R. Vázquez-Lorenzo, "Osserman manifolds in semi-Riemannian geometry", *Lecture Notes in Math.* **1777**, Springer, Berlin, 2002. DOI: 10.1007/b83213. 50, 84, 93

[91] E. García-Río, M. E. Vázquez-Abal, and R. Vázquez-Lorenzo, "Nonsymmetric Osserman pseudo-Riemannian manifolds", *Proc. Amer. Math. Soc.* **126** (1998), 2771–2778. DOI: 10.1090/S0002-9939-98-04666-8. 50, 55, 84, 85

[92] E. García-Río and R. Vázquez-Lorenzo, "Four-dimensional Osserman symmetric spaces", *Geom. Dedicata* **88** (2001), 147–151. DOI: 10.1023/A:1013101719550. 50, 61, 82

[93] H. Geiges, "Contact geometry", *Handbook of Differential geometry, vol. II*, 315–382. Elsevier/North Holland, Amsterdam, 2006. 3

[94] P. Gilkey, "The spectral geometry of a Riemannian manifold", *J. Differential Geom.* **10** (1975), 601–618. http://projecteuclid.org/DPubS?service=UI&version=1.0&verb=Display&handle=euclid.jdg/1214433164 25

[95] P. Gilkey, "Invariance theory, the heat equation, and the Atiyah-Singer index theorem 2nd ed.", *Studies in Advanced Mathematics*, CRC Press, Boca Raton, FL., 1995. 2

[96] P. Gilkey, "Geometric properties of natural operators defined by the Riemann curvature tensor", World Scientific Publishing Co., Inc., River Edge, NJ, 2001. 47, 48, 50, 51, 55

[97] P. Gilkey, "Algebraic curvature tensors which are p Osserman", *Differential Geom. Appl.* **14** (2001), 297–311. DOI: 10.1016/S0926-2245(01)00040-7. 51

[98] P. Gilkey, "The geometry of curvature homogeneous pseudo-Riemannian manifolds". *ICP Advanced Texts in Mathematics*, **2**. Imperial College Press, London, 2007. DOI: 10.1142/p503. 17, 57

[99] P. Gilkey and R. Ivanova, "Spacelike Jordan-Osserman algebraic curvature tensors in the higher signature setting", *Differential Geometry, Valencia, 2001*, 179–186, World Scientific Publishing Co., Inc., River Edge, NJ, 2002. 50

[100] P. Gilkey and R. Ivanova, "The Jordan normal form of Osserman algebraic curvature tensors", *Results Math.* **40** (2001), 192–204. DOI: 10.1007/BF03322705. 50

[101] P. Gilkey, J. V. Leahy, and H. Sadofsky, "Riemannian manifolds whose skew-symmetric curvature operator has constant eigenvalues", *Indiana Univ. Math. J.* **48** (1999), 615–634. DOI: 10.1512/iumj.1999.48.1699. 55

[102] P. Gilkey and S. Nikčević, "Kaehler and para-Kaehler curvature Weyl manifolds", *Publ. Math. Debrecen* **80** (2012), 369–384. 19, 105, 108

[103] P. Gilkey and S. Nikčević, "Kähler–Weyl manifolds of dimension 4", to appear in *Rend. Sem. Mat. Univ. Politec. Torino.* 19, 105, 107, 108

[104] P. Gilkey and S. Nikčević, "4-dimensional (para)-Kähler–Weyl structures", http://arxiv.org/abs/1210.6769. DOI: 10.1007/978-1-4614-4897-6_15. 105

[105] P. Gilkey and S. Nikčević, "(para)-Kähler Weyl structures", Recent trends in Lorentzian Geometry. *Springer Proceedings in Mathematics & Statistics* **26**, 335–353, Springer-Verlag, New York, NY, 2013. DOI: 10.1007/978-1-4614-4897-6_15. 19, 105, 107, 108

[106] P. Gilkey and S. Nikčević, "Affine projective Osserman structures", http://arxiv.org/abs/1304.7482. 57

[107] P. Gilkey, S. Nikčević, and U. Simon, "Geometric theory of equiaffine curvature tensors", *Results Math.* **56** (2009), 275–318. DOI: 10.1007/s00025-009-0438-1. 17

[108] P. Gilkey, S. Nikčević, and U. Simon, "Curvature Properties of Weyl Geometries", *Results Math* **59** (2011), 523–544. DOI: 10.1007/s00025-011-0111-3. 15

[109] P. Gilkey, S. Nikčević, and U. Simon, "Geometric realizations, curvature decompositions, and Weyl manifolds", *J. Geom. Phys.* **61** (2011), 270–275. DOI: 10.1016/j.geomphys.2010.09.022. 15, 16, 17

[110] P. Gilkey, S. Nikčević, and V. Videv, "Manifolds which are Ivanov-Petrova or k-Stanilov", *J. Geom.* **80** (2004), 82–94. DOI: 10.1007/s00022-003-1750-7. 51, 52

[111] P. Gilkey, J.H. Park, and K. Sekigawa, "Universal curvature identities", *Differential Geom. Appl.*, **29** (2011), 770-778. DOI: 10.1016/j.difgeo.2011.08.005. 25

[112] P. Gilkey, J.H. Park, and K. Sekigawa, "Universal curvature identities II", *J. Geom. Phys.* **62**, (2012), 814-825. DOI: 10.1016/j.geomphys.2012.01.002. 25

[113] P. Gilkey and I. Stavrov, "Curvature tensors whose Jacobi or Szabó operator is nilpotent on null vectors", *Bull. London Math. Soc.* **34** (2002), 650–658. DOI: 10.1112/S0024609302001339. 51

[114] P. Gilkey, A. Swann, and L. Vanhecke, "Isoparametric geodesic spheres and a conjecture of Osserman concerning the Jacobi operator", *Quart. J. Math. Oxford Ser. (2)* **46** (1995), 299–320. DOI: 10.1093/qmath/46.3.299. 50, 55, 78

[115] P. Gilkey and T. Zhang, "Algebraic curvature tensors for indefinite metrics whose skew-symmetric curvature operator has constant Jordan normal form", *Houston J. Math.* **28** (2002), 311–328. DOI: 10.1023/A:1015221317388. 52, 55

[116] M. de Gosson, "Symplectic geometry and quantum mechanics", *Operator Theory: Advances and Applications,* **166**. Advances in Partial Differential Equations (Basel). Birkhäuser Verlag, Basel, 2006. 3

[117] A. Gray, "Curvature identities for Hermitian and almost Hermitian manifolds", *Tôhoku Math. J.* **28** (1976), 601–612. DOI: 10.2748/tmj/1178240746. 48

[118] A. Gray and L. Hervella, "The sixteen classes of almost Hermitian manifolds and their linear invariants", *Ann. Mat. Pura Appl. (4)* **123** (1980), 35–58. DOI: 10.1007/BF01796539. xi, 23

[119] A. Gray and L. Vanhecke, "Almost Hermitian manifolds with constant holomorphic sectional curvature", *Časopis Pěst. Mat.* **104** (1979), 170–179. 48

[120] G. S. Hall, "Some remarks on the converse of Weyl's conformal theorem", *J. Geom. Phys.* **60** (2010), 1–7. DOI: 10.1016/j.geomphys.2009.08.002. 27

[121] G. S. Hall, "Covariantly constant tensors and holonomy structure in general relativity", *J. Math. Phys.* **32** (1991), 181–187. DOI: 10.1063/1.529114. 29

[122] N. Hawley, "Constant holomorphic curvature", *Canad. Math. J.* **5** (1953), 53–56. DOI: 10.4153/CJM-1953-007-1. 48

[123] S. Helgason, "Differential geometry and symmetric spaces", *Pure and Applied Mathematics, vol. XII*, Academic Press, New York-London, 1962. 15, 42

[124] U. Hertrich-Jeromin, "Introduction to Möbius differential geometry", *London Mathematical Society Lecture Note Series* **300**, Cambridge University Press, Cambridge, 2003. 27

[125] T. Higa, "Weyl manifolds and Einstein-Weyl manifolds", *Comm. Math. Univ. St. Paul.* **42** (1993), 143-160. xi, 1, 20, 28

[126] T. Higa, "Curvature tensors and curvature conditions in Weyl geometry", *Comm. Math. Univ. St. Paul* **43** (1994), 139–153. xi, 1, 20, 28

[127] N. Hitchin, "Hyperkähler manifolds", Séminaire Bourbaki (1991/92), Ast'erisque **206** (1992), Exp. No. 748, 3, 137–166. 80

[128] W. Hodge, *The theory and applications of harmonic integrals*, Cambridge University Press, Cambridge, 1941. 2

[129] K. Honda and K. Tsukada, "Conformally flat semi-Riemannian manifolds with nilpotent Ricci operators and affine differential geometry", *Ann. Global Anal. Geom.* **25** (2004), 253–275. DOI: 10.1023/B:AGAG.0000023245.73639.93. 11

[130] L. Hörmander, "The Frobenius-Nirenberg theorem", *Ark. Mat.* **5** (1965), 425–432. DOI: 10.1007/BF02591139. 18

[131] J. Igusa, "On the structure of a certain class of Kähler varieties", *Amer. J. Math.* **76** (1954), 669–678. DOI: 10.2307/2372709. 48

[132] M. Itoh, "Affine locally symmetric structures and finiteness theorems for Einstein-Weyl manifolds", *Tokyo J. Math.* **23** (2000), 37–49. DOI: 10.3836/tjm/1255958806. 15

[133] S. Ivanov, "Geometry of quaternionic Kähler connections with torsion", *J. Geom. Phys.* **41** (2002), 235–257. DOI: 10.1016/S0393-0440(01)00058-4. 11

[134] S. Ivanov and I. Petrova, "Riemannian manifold in which the skew-symmetric curvature operator has pointwise constant eigenvalues", *Geom. Dedicata* **70** (1998), 269–282. DOI: 10.1023/A:1005014507809. 52, 56

[135] S. Ivanov and S. Zamkovoy, "Parahermitian and paraquaternionic manifolds", *Differential Geom. Appl.* **23** (2005), 205–234. DOI: 10.1016/j.difgeo.2005.06.002. 49, 79, 83

[136] N. Iwahori, "Some remarks on tensor invariants of $O(n)$, $U(n)$, $Sp(n)$", *J. Math. Soc. Japan* **10** (1958), 146–160. DOI: 10.2969/jmsj/01020145. 24

[137] J. Jost, "Riemannian geometry and geometric analysis", *Universitext*, Springer-Verlag, Berlin, 2002. DOI: 10.1007/978-3-662-04745-3. 4

[138] M. Karoubi, "K-theory, an introduction", *Grundlehren der mathematischen Wissenschaften* **226**, Springer Verlag (Berlin) (1978). 80

[139] J. Kerimo, "AdS pp-waves", *J. High Energy Phys.* (2005), 025, 18 pp. DOI: 10.1088/1126-6708/2005/09/025. 31

[140] S. Kobayashi and K. Nomizu, "Foundations of Differential Geometry vol. I and II", *Wiley Classics Library*. A Wyley-Interscience Publication, John Wiley & Sons, Inc., New York, 1996. 1, 9, 10, 14, 15, 18, 47, 48

[141] G. Kokarev and D. Kotschick, "Fibrations and fundamental groups of Kähler-Weyl manifolds", *Proc. Amer. Math. Soc.* **138** (2010), 997–1010. DOI: 10.1090/S0002-9939-09-10110-7. 106

[142] O. Kowalski, B. Opozda, and Z. Vlášek, "Curvature Homogeneity of affine connections on two-dimensional manifolds", *Colloq. Math.* **81** (1999), 123–139. 60

[143] O. Kowalski, B. Opozda, and Z. Vlášek, "A classification of locally homogeneous affine connections with skew-symmetric Ricci tensor on 2-dimensional manifolds", *Monatsch. Math.* **130** (2000), 109–125. DOI: 10.1007/s006050070041. 56, 60

[144] O. Kowalski, B. Opozda, and Z. Vlášek, "A classification of locally homogeneous connections on 2-dimensional manifolds via group-theoretical approach", *Cent. Eur. J. Math.* **2** (2004), 87–102. DOI: 10.2478/BF02475953. 56, 70

[145] O. Kowalski and M. Sekizawa, "Natural lifts in Riemannian geometry", *Variations, geometry and physics*, 189–207, Nova Sci. Publ., New York, 2009. 7

[146] O. Kowalski and M. Sekizawa, "On natural Riemann extensions", *Publ. Math. Debrecen* **78** (2011), 709–721. 7

[147] S. Lang, "Differential and Riemannian manifolds", *Graduate Texts in Mathematics* **160**, Springer-Verlag, Berlin, 1995. DOI: 10.1007/978-1-4612-4182-9. 18

[148] P. Law and Y. Matsushita, "A spinor approach to Walker geometry", *Comm. Math. Phys.* **282** (2008), 577–623. DOI: 10.1007/s00220-008-0561-y. 29

[149] T. Leistner, "Screen bundles of Lorentzian manifolds and some generalizations of pp-waves", *J. Geom. Phys.* **56** (2006), 2117–2134. DOI: 10.1016/j.geomphys.2005.11.010. 31

[150] T. Leistner, "Conformal holonomy of C-spaces, Ricci-flat, and Lorentzian manifolds", *Differential Geom. Appl.* **24** (2006), 458–478. DOI: 10.1016/j.difgeo.2006.04.008. 29, 31

[151] T. Levi-Civita, "Nozione di parallelismo in una varietà qualunque e consequente specificazione geometrica della curvatura Riemanniana", *Rend. Circ. Mat. Palermo* **42** (1917), 73–205. DOI: 10.1007/BF03014898. xi, 14

[152] Y. Matsushita, "Walker 4-manifolds with proper almost complex structures", *J. Geom. Phys.* **55** (2005), 385–398. DOI: 10.1016/j.geomphys.2004.12.014. 29

[153] P. Matzeu and S. Nikčević, "Linear algebra of curvature tensors on Hermitian manifolds", *An. Stiint. Univ. Al. I. Cuza. Iasi Sect. I. a Mat.* **37** (1991), 71–86. 20

[154] D. McDuff and D. Salamon, "Introduction to symplectic topology", *Oxford Mathematical Monographs*, The Clarendon Press, Oxford University Press, New York, 1998. 3

[155] J. Michelson and X. Wu, "Dynamics of antimembranes in the maximally supersymmetric eleven-dimensional pp wave", *J. High Energy Phys.* (2006), 028, 37 pp. DOI: 10.1088/1126-6708/2006/01/028. 31

[156] H. Mori, "On the decomposition of generalized K-curvature tensor fields", *Tohoku Math. J. (2)* **25** (1973), 225–235. DOI: 10.2748/tmj/1178241382. 28

[157] A. Newlander and L. Nirenberg, "Complex analytic coordinates in almost complex manifolds", *Ann. of Math.* **65** (1957), 391–404. DOI: 10.2307/1970051. 18

[158] A. Nijenhuis and W. Woolf, "Some integration problems in almost-complex and complex manifolds", *Ann. of Math.* **77** (1963), 424–489. DOI: 10.2307/1970126. 18

[159] S. Nikčević, "On the decomposition of curvature tensor fields on Hermitian manifolds", *Differential geometry and its applications (Eger, 1989)*, 555-568, *Colloq. Math. Soc. Janos Bolyai* **56**, North-Holland, Amsterdam 1992. 20

[160] S. Nikčević, "On the decomposition of curvature tensor", *Proceedings of the Ninth Yugoslav Conference on Geometry (Kragujevac, 1992)*. Zb. Rad. (Kragujevac) **16** (1994), 61–68. 20

[161] Y. Nikolayevsky, "Riemannian manifolds whose curvature operator $R(X, Y)$ has constant eigenvalues", *Bull. Austral. Math. Soc.* **70** (2004), 301–319. DOI: 10.1017/S0004972700034523. 56

[162] Y. Nikolayevsky, "Osserman manifolds of dimension 8", *Manuscripta Math.* **115** (2004), 31–53. DOI: 10.1007/s00229-004-0480-y. 50

[163] Y. Nikolayevsky, "Osserman conjecture in dimension $\neq$ 8, 16", *Math. Ann.* **331** (2005), 505–522. DOI: 10.1007/s00208-004-0580-8. 50

[164] Y. Nikolayevsky, "Conformally Osserman manifolds", *Pacific J. Math.* **245** (2010), 315–358. DOI: 10.2140/pjm.2010.245.315. 51

[165] Y. Nikolayevsky, "Conformally Osserman manifolds of dimension 16 and a Weyl-Schouten theorem for rank-one symmetric spaces", *Ann. Mat. Pura Appl.* **191** (2012), 677–709. DOI: 10.1007/s10231-011-0201-6. 51

[166] K. Nomizu, "Introduction to affine differential geometry, Part I", *MPI Preprint 88-37*, 1988. 13, 39

[167] K. Nomizu, "On the decomposition of generalized curvature tensor fields, Codazzi, Ricci, Bianchi and Weyl revisited", *Differential geometry (in honor of K. Yano)*, 335-345. Kinokuniya, Tokyo, 1972. 17

[168] K. Nomizu and T. Sasaki, "Affine differential geometry", *Cambridge Tracts in Mathematics* **111**, Cambridge Univ. Press, Cambridge, 1993. 11

[169] B. O'Neill, "Semi-Riemannian geometry. With applications to relativity", *Pure and Applied Mathematics* **103**, Academic Press, Inc., New York, 1983. 15, 44

[170] Z. Olszak, "On conformally recurrent manifolds II. Riemann extensions", *Tensor (N.S.)* **49** (1990), 24–31. 29

[171] Z. Olszak, "On the existence of generalized complex space forms", *Israel J. Math.* **65** (1989), 214–218. DOI: 10.1007/BF02764861. 48, 49

[172] B. Opozda, A classification of locally homogeneous connections on 2-dimensional manifolds, *Differential Geom. Appl.* **21** (2004), 173–198. DOI: 10.1016/j.difgeo.2004.03.005. 56, 62

[173] R. Osserman, "Curvature in the eighties", *Amer. Math. Monthly* **97** (1990), 731–756. DOI: 10.2307/2324577. 49, 55

[174] G. Ovando, "Invariant pseudo-Kähler metrics in dimension four", *J. Lie Theory* **16** (2006), 371–391. 16

[175] A. Özdeğer, "On sectional curvatures of a Weyl manifold", *Proc. Japan Acad. Ser. A Math. Sci.* **82** (2006), 123–125. DOI: 10.3792/pjaa.82.123. 16

[176] J. Patera, R. T. Sharp, P. Winternitz, and H. Zassenhaus, "Invariants of real low dimensional Lie algebras", *J. Math. Phys.* **17** (1976), 986–994. DOI: 10.1063/1.522992. 117

[177] E. M. Patterson and A. G. Walker, "Riemann extensions", *Quart. J. Math., Oxford Ser. (2)* **3** (1952), 19–28. DOI: 10.1093/qmath/3.1.19. xi, 29, 33

[178] H. Pedersen, Y. Poon, and A. Swann, "The Einstein-Weyl equations in complex and quaternionic geometry", *Differential Geom. Appl.* **3** (1993), 309–321. DOI: 10.1016/0926-2245(93)90009-P. 106

[179] H. Pedersen and A. Swann, "Riemannian submersions, four manifolds, and Einstein-Weyl geometry", *Proc. London Math. Soc.* **66** (1991), 381–399. DOI: 10.1112/plms/s3-66.2.381. 15

[180] H. Pedersen and K. Tod, "Three-dimensional Einstein-Weyl geometry", *Adv. Math.* **97** (1993), 74–109. DOI: 10.1006/aima.1993.1002. 15

[181] V. Perlick, "Observer fields in Weylian space time models", *Classical Quantum Gravity* **8** (1991), 1369–1385. DOI: 10.1088/0264-9381/8/7/013. 15

[182] V. Pravda, A. Pravdová, A. Coley, and R. Milson, "All spacetimes with vanishing curvature invariants", *Classical Quantum Gravity* **19** (2002), 6213–6236. DOI: 10.1088/0264-9381/19/23/318. 25

[183] F. Prüfer, F. Tricerri, and L. Vanhecke, "Curvature invariants, differential operators and local homogeneity", *Trans. Amer. Math. Soc.* **348** (1996), 4643–4652. DOI: 10.1090/S0002-9947-96-01686-8. 25

[184] G. de Rham, "Sur la reductibilité d'un espace de Riemann", *Comment. Math. Helv.* **26** (1952), 328–344. DOI: 10.1007/BF02564308. 30

[185] G. de Rham, "La théorie des formes différentielles extérieures et l'homologie des variétés différentiables", *Rend. Mat. e Appl. (5)* **20** (1961), 105–146. DOI: 10.1007/978-3-642-10952-2_1. 2

[186] G. Ricci, "Sulla derivazione covariante ad una forma quadratica differenziale", *Rom. Acc. L. Rend.* **3** (1887), 15–18. 8

[187] G. Ricci, "Direzioni e invarianti principali in una varietà qualunque", *Ven. Ist. Atti* **63** (1903–1904), 1233–1239. 10

[188] G. Ricci and T. Levi-Civita, "Méthodes de calcul différential absolu et leurs applications", *Math. Ann.* **54** (1900), 125–201. DOI: 10.1007/BF01454201. xi, 8, 9

[189] B. Riemann, "On the hypotheses which lie at the foundation of geometry", *Nature VIII. Nos. 183, 184* (1873), 14–17, 36, 37. (Translated by W. Clifford). 9

[190] P. A. Schirokow and A. P. Schirokow, *Affine Differentialgeometrie*, Deutsche Übersetzung: Olaf Neumann; Wissenschaftliche Redaktion: Hans Reichardt B. G. Teubner Verlagsgesellschaft, Leipzig, 1962. 11, 12

[191] J. A. Schouten, "Über die konforme Abbildung n-dimensionaler Mannigfaltigkeiten mit quadratischer Maßbestimmung auf eine Mannigfaltigkeit mit euklidischer Maßbestimmung", *Math. Z.* **11** (1921), 58–88. DOI: 10.1007/BF01203193. 27

[192] U. Simon, "Affine differential geometry", *Handbook of Differential Geometry, vol. I*, 905–961, North-Holland, Amsterdam, 2000. DOI: 10.1016/S1874-5741(00)80012-6. 11

[193] U. Simon, A. Schwenck-Schellschmidt, and H. Viesel, "Introduction to the affine differential geometry of hypersurfaces", *Lecture Notes Science University Tokyo*, 1991. 11

[194] I. M. Singer and J. A. Thorpe, "The curvature of 4-dimensional Einstein spaces", *Global Analysis (Papers in honor of K. Kodaira)*, 355–365. Univ. Tokyo Press, Tokyo, 1969. xi, 1, 20, 27, 28

[195] M. Sitaramayya, "Curvature tensors in Kaehler manifolds", *Trans. Amer. Math. Soc.* **183** (1973), 341–353. DOI: 10.1090/S0002-9947-1973-0322722-1. 28

[196] M. Spivak, "A comprehensive introduction to differential geometry", Second edition. Publish or Perish, Inc., Wilmington, Del., 1979. 4, 14

[197] G. Stanilov and V. Videv, "On a generalization of the Jacobi operator in the Riemannian geometry", *Annuaire Univ. Sofia Fac. Math. Inform.* **86** (1992), 27–34. 51

[198] R. Strichartz, "Linear algebra of curvature tensors and their covariant derivatives", *Canad. J. Math.* **40** (1988), 1105–1143. DOI: 10.4153/CJM-1988-046-7. 20

[199] K. Tod, "Compact 3 dimensional Einstein-Weyl structures", *J. London Math. Soc.* **45** (1992), 341–351. DOI: 10.1112/jlms/s2-45.2.341. 15

[200] F. Tricerri and L. Vanhecke, "Curvature tensors on almost Hermitian manifolds", *Trans. Amer. Math. Soc.* **267** (1981), 365–397. DOI: 10.1090/S0002-9947-1981-0626479-0. xi, 1, 20, 28, 29, 37, 125

[201] I. Vaisman, "Generalized Hopf manifolds", *Geom. Dedicata* **13** (1982), 231–255. DOI: 10.1007/BF00148231. 106

[202] I. Vaisman, "A survey of generalized Hopf manifolds", Conference on differential geometry on homogeneous spaces (Turin, 1983). *Rend. Sem. Mat. Univ. Politec. Torino* **1983**, Special Issue, 205–221 (1984). 106

[203] L. Vanhecke and T. J. Willmore, "Riemann extensions of D'Atri spaces", *Tensor (N.S.)* **38** (1982), 154–158. DOI: 10.1007/978-1-4612-2432-7_9. 29

[204] A. G. Walker, "Canonical form for a Riemannian space with a parallel field of null planes", *Quart. J. Math., Oxford Ser. (2)* **1** (1950), 69–79. DOI: 10.1093/qmath/1.1.69. 31

[205] A. Weinstein, "Symplectic geometry", *Bull. Amer. Math. Soc.* **5** (1981), 1–13. DOI: 10.1090/S0273-0979-1981-14911-9. 3

[206] R. Wells, "Differential analysis on complex manifolds", *Graduate Texts in Mathematics* **65**, Springer-Verlag, New York-Berlin, 1980. DOI: 10.1007/978-1-4757-3946-6. 18

[207] H. Weyl, "Reine Infinitesimalgeometrie", *Math. Z.* **2** (1918), 384–411. DOI: 10.1007/BF01199420. 27

[208] H. Weyl, "Zur Infinitesimalgeometrie: Einordnung der projektiven und der konformen Auffassung", *Gött. Nachr.* (1921), 99–112. 27, 34

[209] H. Weyl, *Space-time matter.* Fourth edition, Dover Publ., New York, 1951. 15

[210] H. Weyl, "The Classical Groups. Their Invariants and Representations", *Princeton University Press*, Princeton, NJ, 1939. 24

[211] T. J. Willmore, "Riemann extensions and affine differential geometry", *Results Math.* **13** (1988), 403–408. DOI: 10.1007/BF03323255. 29, 34

[212] J. Wolf, "Spaces of constant curvature. Sixth edition", *AMS Chelsea Publishing*, Providence, RI, 2011. 15, 45

[213] Y. C. Wong, "Two dimensional linear connexions with zero torsion and recurrent curvature", *Monatsh. Math.* **68** (1964), 175–184. DOI: 10.1007/BF01307120. 39, 40, 60, 61, 62

[214] H. Wu, "On the de Rham decomposition theorem", *Illinois J. Math.* **8** (1964), 291–311. http://projecteuclid.org/DPubS?service=UI&version=1.0&verb= Display&handle=euclid.ijm/1256059674 30

[215] K. Yano and S. Ishihara, "Tangent and cotangent bundles", *Pure and Applied Mathematics* **16**, Marcel Dekker Inc., New York, 1973. 4, 5, 6, 34

[216] T. Zhang, "Manifolds with indefinite metrics whose skew-symmetric curvature operator has constant eigenvalues", Steps in differential geometry (Debrecen, 2000) 401–407, Inst. Math. Inform. Debrecen 2001. 52

Authors' Biographies

EDUARDO GARCÍA-RÍO

Eduardo García-Río[1] is a Professor of Mathematics and a member of the Institute of Mathematics of the University of Santiago de Compostela (Spain). He received his Ph.D. degree in 1992 from the University of Santiago de Compostela and is a member of the editorial board of the *Journal of Geometric Analysis*. His research specialities are Differential Geometry and Mathematical Physics.

PETER GILKEY

Peter Gilkey[2] is a Professor of Mathematics and a member of the Institute of Theoretical Science at the University of Oregon (USA). He is a fellow of the American Mathematical Society and is a member of the editorial board of *Results in Mathematics, Differential Geometry and Applications,* and the *International Journal of Geometric Methods to Mathematical Physics*. He received his Ph.D. in 1972 from Harvard University under the direction of L. Nirenberg. His research specialties are Differential Geometry, Elliptic Partial Differential Equations, and Algebraic topology. He has published more than 230 research articles and books.

[1] Department of Geometry and Topology, Faculty of Mathematics, University of Santiago de Compostela, 15782 Santiago de Compostela, Spain.
email: eduardo.garcia.rio@usc.es
[2] Mathematics Department, University of Oregon, Eugene OR 97403 USA
email: gilkey@uoregon.edu

STANA NIKČEVIĆ

Stana Nikčević[3] is a Professor of Mathematics at the University of Belgrade (Serbia). She received her Ph.D. from the University of Belgrade at the Mathematical Faculty and has been working at the University since 1974. During this period, she also sporadically worked at the University of Banja Luka (Bosnia and Hercegovina) and at the Mathematical Faculty in Kragujevac (Serbia). Her research mainly focuses on Differential Geometry. She has maintained international cooperation and has gone on short visits to the TU Berlin, Charles University Prague, Universitate Pierre et Marie Curie (Paris VI), University of Oregon (USA), and University of Santiago de Compostela (Spain).

RAMÓN VÁZQUEZ-LORENZO

Ramón Vázquez-Lorenzo[4] is a member of the research group in Riemannian Geometry at the Department of Geometry and Topology of the University of Santiago de Compostela (Spain). He is a member of the Spanish Research Network on Relativity and Gravitation. He received his Ph.D. in 1997 from the University of Santiago de Compostela. His research focuses mainly on Differential Geometry with special emphasis on the study of the curvature and the algebraic properties of curvature operators in the Lorentzian and in higher signature settings. He has published more than 45 research articles and books.

[3]Mathematical Institute, Sanu, Knez Mihailova 36, p.p. 367
11001 Beograd, Serbia
email: stanan@mi.sanu.ac.rs
[4]Department of Geometry and Topology, Faculty of Mathematics, University of Santiago de Compostela,
15782 Santiago de Compostela, Spain.
email: ravazlor@edu.xunta.es

Index

Printed in the United States
by Baker & Taylor Publisher Services